Group Theory and Its
Physical Applications

CHICAGO LECTURES IN PHYSICS

Group Theory and Its Physical Applications

L. M. FALICOV

Notes compiled by A. LUEHRMANN

Chicago & London
THE UNIVERSITY OF CHICAGO PRESS

Library of Congress Catalog Card Number: 66-13867

THE UNIVERSITY OF CHICAGO PRESS, CHICAGO 60637
THE UNIVERSITY OF CHICAGO PRESS, LTD., LONDON W.C. 1.

Preface

These notes constitute the subject matter of a graduate course offered by the Physics Department of the University of Chicago during the summer quarter of 1964. They were taken from the lectures and transcribed by Arthur Luehrmann with the collaboration of Dr. P. M. Lee. They do not comprise a complete book or a significantly new research contribution to the field; they simply contain a selection of topics in the theory of groups and its application to physical problems, topics that strongly reflect the point of view and personal taste of the author. For more complete treatments and for further applications, which were necessarily left out of a one-quarter course, the reader is referred to the many good books already available in print, in particular those of Wigner, Heine, and Tinkham.

I would like to show my appreciation to Mr. A. Luehrmann, who contributed to this book not only his perseverance and his work, but many new ideas regarding presentation and organization of the subject matter. I am also very grateful to Miss Dianne Berger for her excellent typing of the manuscript and to Mr. George Tzouras for the proofreading and organization of this final version.

L. M. F.

Contents

I

Fundamental Definitions and Theorems

SETS AND OPERATIONS

A <u>set</u> E is a collection of entities a, b, c, ...
called <u>elements</u>. The number of elements may be zero (the
<u>null</u> set), finite, denumerably infinite, or non-denumera-
bly infinite.

Examples:

(a) Finite: $\{1,-1\}; \{1,-1,i,-i\}; \{1, -\frac{1}{2}+i\frac{\sqrt{3}}{2}, -\frac{1}{2}-i\frac{\sqrt{3}}{2}\}$.

(b) Denumerably infinite: Natural Numbers: 1, 2,
3, 4, ... Integers: 0, ±1, ±2, ...
Rational numbers between 0 and 1: 1, $\frac{1}{2}$, $\frac{1}{3}$, $\frac{1}{4}$,
$\frac{2}{3}$, ...

(c) Non-denumerably infinite: Set of all real
numbers; set of all complex numbers.

(d) Operations may also form sets. Thus the six
operations which transform the combination
(ABC) into {(ABC), (BCA), (CAB), (BAC), (ACB),
(CBA)} may be taken as elements of a set.

<u>Operations</u> may be defined between two elements of
a set. We may symbolize this by writing ab=c or ba=d. In
the first case we say that "a operating on b gives c," and
in the second case, "b operating on a gives d."

Definition: If an operation ab=c is defined for all
elements a and all elements b of a set E, and if c is always in E, then the operation is called an <u>internal</u> <u>opera-</u>
<u>tion</u>.

Examples: Let E be the set of natural numbers;

(a) addition $\qquad$ 1 + 2 = 3,⎫
(b) multiplication 1 x 2 = 2,⎬ internal operations
(c) exponentiation 2^3= 8, ⎭

(a') subtraction 1 - 2 = -1 ⎸

(b') division $\qquad 1 \div 2 = \frac{1}{2}$ $\left.\vphantom{\begin{array}{c}a\\b\end{array}}\right\}$ not internal
(c') n^{th} root $\qquad \sqrt[3]{2} \quad = \sqrt[3]{2}$

Let E be the set of all 3-component vectors.

(a) Addition and vector product are internal operations.

(b) Scalar product is not internal.

Further Definitions:

(1) Associativity. A set E is said to be associative under a given operation ab = q if for all a, b, and c, (ab)c = a(bc).

(2) Commutativity. A set E is said to be commutative under a given operation ab = q if for all a and b, ab = ba.

(3) Regular element. a is a regular element of E under the operation ab = c, if for every element x and y in E, the equality ax = ay implies x = y.

Example: Zero is a regular element of the set of integers under the operation of addition, but not under the operation of multiplication.

(4) Unit elements. If a set E contains an element e (or f) such that for all a in E, ae = a (or fa = a), then e (or f) is called the right (or left) unit element of E under the operation in question.

(5) Symmetric elements. If a set E contains an element a_R^{-1} (or a_L^{-1}) such that for a given element a in E, $aa_R^{-1} = e$ (or $a_L^{-1}a = f$), then a_R^{-1} (or a_L^{-1}) is called the right (left) symmetric element of a under the given operation.

Theorems:

I. If a and b are regular elements of E under an associative operation, then ab is also regular.

Proof: Let x be any element of E and let y be an element of E for which (ab)x = (ab)y. Then a(bx) = a(by). But since a is regular, bx = by, and since b is regular, x = y. Hence ab is regular.

II. If both unit elements exist, then they are equal.

Proof: Apply the left unit element f to e. Then by definition (4), fe = e. Similarly apply the right unit element e to f: fe = f. Hence e = f.

III. If E is associative under a given operation and if a, a_L^{-1}, and a_R^{-1} exist, then a_L^{-1} and a_R^{-1} are equal.

Proof: $a_R^{-1} = fa_R^{-1} = (a_L^{-1}a)a_R^{-1} = a_L^{-1}(aa_R^{-1}) = a_L^{-1}e = a_L^{-1}.$

IV. Corollary to III. Subject to the same conditions, a is regular.

Proof: Let x be any element of E and y also be an element of E such that ax = ay. Then $a^{-1}(ax) = a^{-1}(ay)$ and $(a^{-1}a)x = (a^{-1}a)y$. Hence x = y and a is regular.

V. Corollary to III. Subject to the same conditions, the symmetric element of ab, $(ab)^{-1}$, is equal to $b^{-1}a^{-1}$.

Proof: $(b^{-1}a^{-1})(ab) = b^{-1}(a^{-1}a)b = b^{-1}eb = b^{-1}b = e.$

APPLICATIONS

Definition of the concept of Application

If two sets E and E' with elements {x}, {x'} exist such that every element of E' is related to every element of E, x' = f(x), then E' is related to E by application. Note that f(x) is single-valued but its inverse is not in general single-valued. That is, we are dealing with a many-to-one transformation and not in general with a one-to-one transformation.

Let a' = f(a), b' = f(b), c' = f(c). Consider an operation defined on the elements of E, c = ab. If there exists a (generally different) operation on E' such that c' = a'b', whenever c = ab and for all a and b, then the application is called regular.

Example: Let E be the set of positive real numbers

and E' the set of all real numbers. Let $x' = \log x$. The operation in E is ordinary multiplication. If we take the operation in E' to be addition, we see that $c = a \times b$ implies $c' = a' + b'$ and E' is related to E by a <u>regular application</u>.

Types of regular application

(1) <u>Isomorphic</u>. The relation between x and x' is one-to-one. In other words, the inverse of $f(x)$ is single-valued. The above example $x' = \log x$ is an isomorphism. The natural numbers are isomorphic with the even numbers, $x' = 2x$, when the operation in both sets is addition.

(2) <u>Automorphic</u>. E and E' are isomorphic under the same operation, and the elements of E and E' are identical.

Example: Write $\omega = -\frac{1}{2} + \frac{i\sqrt{3}}{2}$, $E = \{1, \omega, \omega^2\}$, E' is given by $x' = x^2$, $E' = \{1, \omega^2, \omega\}$, since $(\omega^2)^2 = \omega^4 = \omega$. If the operation is ordinary multiplication, it is easy to confirm that $c = ab$ implies $c' = a'b'$ and the application is regular.

(3) <u>Homomorphic</u>. This is the general case of a regular application

Example: $E = \{0, 1, 2, \dots \infty\}$, $E' = \{0, 1, 2, 3, 4\}$. E' is homomorphic with E if the operation is addition in E and addition modulo 5 in E' or multiplication in E and multiplication modulo 5 in E'.

If E and E' are isomorphic, then they possess in common any of the properties 1 - 5 that either possesses. If they are homomorphic, E' has the properties of E but the reverse is not necessarily true. Thus in the last example each element of E' has a symmetric element, but this is not true of E.

EQUIVALENCE

The essential properties of the equivalence relationship, indicated by the symbol $\equiv$, are the following:

(1) $a \equiv a$ (reflexive)

(2) If $a \equiv b$, then $b \equiv a$ (symmetric)

(3) If $a \equiv c$ and $b \equiv c$, then $a \equiv b$ (transitive).

Two elements of a set are said to be equivalent if they have some property in common.

Factored Set

Definition: C(a) is a subset of E containing all elements equivalent to a. C(b) is similarly defined.

Theorem VI: Either C(a) is identical to C(b) or they have no element in common.

Proof: Let c belong to C(a) and C(b). Then $a \equiv c$ and $b \equiv c$. Hence $a \equiv b$, and all elements equivalent to b are also equivalent to a. Hence $C(a) = C(b)$.

Now we have a way of decomposing a set E into unique subsets, C(a), C(b), C(c)... . The set of subsets is called the factored set, and each subset is itself an element of the factored set. An operation between the new elements can be defined in terms of the operation between the original elements.

Definition: $C(a) \cdot C(b) = C(a.b)$.

This definition is not unique since even if $a \equiv a'$, $b \equiv b'$ whence $C(a) = C(a')$, $C(b) = C(b')$. It does not follow that $ab \equiv a'b'$. Hence $C(ab) \neq C(a'b')$ in general. To make the definition unique we must have regular equivalence within each subset.

Definition: a is said to be regularly equivalent to a' if their ordinary equivalence implies that for all x in E, $ax \equiv a'x$ and $xa \equiv xa'$. Now we prove that with this condition C(ab) is unique.

$$a \equiv a' \Rightarrow ab \equiv a'b \qquad \Rightarrow ab \equiv a'b'$$
$$b \equiv b' \Rightarrow a'b \equiv a'b'$$

Hence $\qquad C(ab) = C(a'b')$.

GROUPS

Definition: A group is a set of elements for which an operation is defined with the following four properties.

(1) Closure. For every element a, b, in the group G, the result of ab is also in G. In other words, the operation is an internal one.

(2) <u>Associativity</u>. For all a, b, c, in G, (ab)c
= a(bc).

(3) <u>Unit Element</u>. For every a in G, there exists
some element e such that ea = a.

(4) <u>Inverse Element</u>. For every a in G, there
exists a corresponding element a^{-1} such that $a^{-1}a$ = e.
From (2), (3), and (4), the following properties can be
easily derived.

(5) <u>Cancellation Law</u>. All elements are regular
on the left, i.e.,

$$ax = ay \Rightarrow x = y$$

for all a in G.

<u>Proof</u>: $ax = ay \Rightarrow a^{-1}(ax) = a^{-1}(ay)$
$\Rightarrow (a^{-1}a)x = (a^{-1}a)y \Rightarrow ex = ey \Rightarrow x = y.$

(3') <u>Unit Element on the Right</u>.

$$ae = a = ea$$

<u>Proof</u>: $a^{-1}(ae) = (a^{-1}a)e = ee = e = a^{-1}a$
$a^{-1}(ae) = a^{-1}a \Rightarrow ae = a = ea$

(4') <u>Inverse Element on the Right</u>.

$$aa^{-1} = e = a^{-1}a$$

<u>Proof</u>: $a^{-1}(aa^{-1}) = (a^{-1}a)a^{-1} = ea^{-1} = a^{-1}e$
$\Rightarrow aa^{-1} = e = a^{-1}a$

(5') <u>Cancellation Law</u>. All elements are regular
on the right. The proof is identical to (5), using (3')
and (4') instead of (3) and (4).

Properties (1), (2), (3) and (4) constitute the general definition of a group. If in addition the group has
the further property of <u>commutativity</u> (ab = ba for all a,
b, in G), then G is said to be <u>Abelian</u>.

<u>Examples of groups</u>

(1) Integers under addition. e = 0, a^{-1}= -a. This is
an infinite abelian group.

(2) Rational numbers excluding zero under multiplica-

tion. e = 1, a^{-1} = 1/a. This is also infinite abelian.

(3) The set 1, -1, i, -i under multiplication. e = 1;
$(-1)^{-1}$ = (-1); $(i)^{-1}$ = -i; $(-i)^{-1}$ = i. Finite abelian.

(4) The set of all non-singular n × n matrices under matrix multiplication. This is infinite non-abelian.

$$e = \begin{pmatrix} 1 & 0 & 0 & \cdots \\ 0 & 1 & 0 & \cdots \\ 0 & 0 & 1 & \cdots \\ \vdots & \vdots & \vdots & \end{pmatrix}$$

(5) The set of permutation operations which takes ABC into {ABC, BCA, CAB, ACB, CBA, BAC}. Thus

$$e = \begin{pmatrix} ABC \\ ABC \end{pmatrix}; \quad \alpha = \begin{pmatrix} ABC \\ BCA \end{pmatrix}; \quad \beta = \begin{pmatrix} ABC \\ CAB \end{pmatrix}; \quad \lambda = \begin{pmatrix} ABC \\ ACB \end{pmatrix}; \quad \mu = \begin{pmatrix} ABC \\ CBA \end{pmatrix}; \quad \nu = \begin{pmatrix} ABC \\ BAC \end{pmatrix}.$$

The operation $\alpha\beta$ is defined to mean: "first do permutation β, then do permutation α on the previous result". It is easy to see that $\alpha\beta$ is equivalent to e. Thus we may derive the entire group multiplication table shown

1st / 2nd	e	α	β	λ	μ	ν
e	e	α	β	λ	μ	ν
α	α	β	e	ν	λ	μ
β	β	e	α	μ	ν	λ
λ	λ	μ	ν	e	α	β
μ	μ	ν	λ	β	e	α
ν	ν	λ	μ	α	β	e

This contains all the information on the group. This group is non-abelian, e.g., $\mu\alpha \neq \alpha\mu$. Mathematically it is the group multiplication table which characterizes any group. All groups with the same group table are mathematically identical as regards their group properties. That is, they are all isomorphic with one another.

Properties of the Group Table

(1) A given element appears in a given row (or column) once and once only.

Proof: Consider the row of element a. Let b_1 and b_2 be elements such that $ab_1 = c$, $ab_2 = c$. Multiply from the left by a^{-1} in both equations. $b_1 = a^{-1}c$, $b_2 = a^{-1}c$. By closure, $a^{-1}c$ is in G; hence $b_1 = b_2$. That is b_1 and b_2 cannot be different elements and both give element c when multiplied by a. Since there are as many elements in a row (or column) as there are elements in G, and since each element can appear once at most, all the elements must appear at least once.

(2) Every row (or column) is different from every other row (or every other column). This follows directly from (1).

(3) For an abelian group, the table is symmetric across the principal diagonal.

Cyclic Groups

If an entire group of n elements may be generated from one element, the group is said to be cyclic. It may be written in terms of the generating element a as $\{a, a^2, a^3, \ldots a^n\}$, where one of these elements is the unit element. In fact $a^n = e$, since if $a^m = e$ for m<n, then $a^n = a^{n-m}a^m = a^{n-m}$. That is a^n would not be different from earlier elements a^{n-m}. n is said to be the order of the group. Every cyclic group is abelian but not conversely.

Order of an element

Consider the set of all elements which can be generated from a in G: $\{a, a^2, \ldots a^n\}$. If this set is finite, then $a^n = e$ and n is said to be the order of the element a. This smaller set is also a group - in fact a cyclic group. In the permutation group example given before, α and β are third order elements, and λ, μ and ν are second order elements.

COMMON GROUPS AND STANDARD SYMBOLS

	Symbol	Elements	Operation	Unit Elem.	Inverse Elem. to a	Properties
	Z	Integers	Addition	0	$-a$	Abelian
	Z_n	Remainder $\left(\frac{m}{n}\right)$	Addition modulo n	0	$-a+n$	Cyclic
	Q	Rational numbers	Addition	0	$-a$	Abelian
	Q^*	Rational numbers except zero	Mult.	1	$1/a$	Abelian
	R	Real numbers	Addition	0	$-a$	Abelian
	R^*	Real numbers except zero	Mult.	1	$1/a$	Abelian
	C	Complex numbers	Addition	0	$-a$	Abelian
	C^*	Complex numbers except zero	Mult.	1	$\frac{1}{a}=\frac{a^*}{|a|^2}$	Abelian
	S	Complex numbers on unit circle	Mult.	1	$\frac{1}{a}=a^*$	Abelian
	S_n	n^{th} roots of unity	Mult.	1	$\frac{1}{a}=a^*$	Isomorphic to Z_n
General linear	$GL(n)$	$n \times n$ matrices	Matrix Mult.	Unit matrix	a^{-1}	
Gen. linear real	$GL(nR)$	$n \times n$ real matrices	"	"	a^{-1}	
Special linear	$SL(n)$	nxn matrices, determinant=1	"	"	a^{-1}	
Unitary	$U(n)$	$n \times n$ unitary matrices	"	"	$a^\dagger$	
Special unitary	$SU(n)$	nxn unitary matrices, determinant=1	"	"	$a^\dagger$	
Orthogonal	$O(n)$	$n \times n$ real unitary matrices	"	"	$\tilde{a}$	

Properties of Finite Groups

(1) Every element has a finite order $n (a^n = e)$. e is a first order element. All others are higher than first.

(2) _Rearrangement Theorem_. Multiplication of each element a_i in G by one element b in G, $a_i b = a_i'$ reproduces the group G but generally in a different order. This was proved as a property of the multiplication table.

Examples of finite groups:

(1) Order 1.   e.

(2) Order 2.

	e	a
e	e	a
a	a	e

Cyclic, isomorphic to S_2 (1, -1) and to $Z_2 (0,1)$.

(3) Order 3.

	e	a	b
e	e	a	b
a	a	b	e
b	b	e	a

Cyclic, isomorphic to S_3, Z_3.

(4) Order 4.

	e	a	b	c
e	e	a	b	c
a	a	b	c	e
b	b	c	e	a
c	c	e	a	b

Cyclic, isomorphic to S_4, Z_4.

(5) Order 4.

	e	a	b	c
e	e	a	b	c
a	a	e	c	b
b	b	c	e	a
c	c	b	a	e

Non-cyclic since $x^2 = e$, but abelian.

(6) Order 5.

Cyclic, isomorphic to S_5, Z_5.

(7) Order 6.

There is one cyclic group isomorphic to S_6, Z_6 and there is a non-abelian group isomorphic to the permutation group given as an example on page 7. This is the lowest order non-abelian group.

SUBGROUPS, COMPLEXES, COSETS, AND CLASSES

Definition of a Subgroup:

A set S is said to be a subgroup of G if

    (1)   all a in S are also in G,

    (2)   for any a and b in S, then ab is in S,

    (3)   all a in S satisfy the four group postulates.

Properties (2) and (3) may be put in an equivalent way as follows. The set S is chosen from G in such a way that whenever any a and any b are in S then ab^{-1} is also in S. In particular S must contain e, because when b = a, aa^{-1} = e must be in S. Now if S contains e and any element b, it must contain b^{-1}.

    Proof: Put a = e. Then $eb^{-1} = b^{-1}$ must be in S. Finally, since S contains b^{-1} as well as b, it follows that $a(b^{-1})^{-1} = ab$ is in S.

Complexes

    A complex is a set of elements from a group G considered as a whole. It will be labeled with a capital letter. If A is a complex and a is an element of G, then aA is the complex of left products of a with elements of A. Similarly, AB is the left product of every element of A with every element of B, where we count only once any product which occurs more than once.

Cosets

    Let S be a subgroup of G. If x is an element of G not in S, then the complex Sx is called a right coset of S and xS is called a left coset of S. Cosets are never groups because they do not contain e. If another element y of G is neither in S nor Sx, then the coset Sy has no element in common with either S or Sx, as Sx has no element in common with S.

    Proof: Let s_i be any element of S, and let $s_i x = w$. Multiply on the left by s_i^{-1}; $x = s_i^{-1} w$. But $s_i^{-1} = s_j$ another element of S. Now if w is in S, then so is $s_j w$. But by hypothesis, x is not in S. So w must not be in S. The proof of the next step is similar. Let y be an element

of G not in S or Sx. It immediately follows that Sy is
not in S. We prove further that Sy is not in Sx. Suppose
on the contrary that there exists some element q common to
both Sx and Sy. That is, $s_i x = q$, $s_j y = q$. Then $s_i x = s_j y$,
$y = s_j^{-1} s_i x$. But since S is a subgroup $s_j^{-1} s_i$ is some s_k.
Hence $y = s_k x$. That is y is in Sx contrary to hypothesis.
Hence no such q exists.

Consequently a finite group can be written as a
factored set, $G = S + Sx + Sy + ...$, with a finite number
of cosets. The factorization is not unique, depending as
it does on the choice of S and x, y, ... etc.

Important Theorem

Let h = number of elements of G and g = number of
elements of some subgroup S. Then $\frac{h}{g}$ = integer.

Proof: Let n = number of terms in factored set. If
there are g elements in S, then there are g different elements in Sx. Similarly for Sy etc., until G is exhausted.
Hence h = ng. Consequences are

(1) If h is prime, the only subgroups are G and e
(2) If h is prime, G is cyclic.

Proof: Suppose G non-cyclic. Then there exists an
element b such that for no a ≠ b in G and for no m does
$a^m = b$. But for a ≠ e, the set $\{a, a^2, ..., a^h\}$ is a
subgroup of order between 2 and h-1. This contradicts (1)
above.

Definition of Class

A complex of elements T is said to form a class if
for every element x in G, $xTx^{-1} = T$. The elements of T
are equivalent; that is, they are related by an equivalence
which satisfies reflexivity, transitivity, and symmetry
properties. The unit element e is obviously in a class by
itself. It is possible to express G as a new sort of factored set, $G = e + T_1 + T_2 + ...$. If G is abelian, every
element is in a class by itself. In finding the members
of the class of a, we look at all products, $xax^{-1} = axx^{-1} =$
$ae = a$. Hence no new element may be found.

Note: Generally a physically significant character-istic may be ascribed to each class. Thus in the permutation group example, the even and odd permutations fall into separate classes. In the group of symmetry transformations of the equilateral triangle, the rotations and reflections fall into separate classes. In a vague sort of way this is understandable, since the definition of a class implies an invariance property in the sense that x^{-1} undoes whatever x does. One might say that multiplication by 2 is quite different from multiplication by 3; whereas rotations about equivalent symmetry axes are not so different.

Invariant Subgroup

Let x be an element of G but not necessarily of subgroup S. First we prove that the complex $S' = xSx^{-1}$ is also a subgroup and that it is isomorphic with S.

(a) No two elements in S' are the same.

Proof: Let $s_i' = xs_ix^{-1}$ and $s_j' = xs_jx^{-1}$. Then $xs_ix^{-1} = xs_jx^{-1}$, and $x^{-1}xs_ix^{-1}x = x^{-1}xs_jx^{-1}x$ or $es_ie = es_je$, $s_i = s_j$. We assume that all elements of S are distinct; so our original supposition was wrong. Hence there is a one-to-one correspondence between S and S'.

(b) If $s_is_j = s_k$, then $s_i's_j' = s_k'$.

Proof: $s_i's_j' = xs_ix^{-1}xs_jx^{-1} = xs_ies_jx^{-1} = xs_is_jx^{-1} = xs_kx^{-1} = s_k'$. Hence the group tables for S and S' are identical. This also proves closure.

(c) Unit element is $e' = x^{-1}ex = x^{-1}x = e$.

(d) Inverse of s' is $s^{-1}{}'$. $s^{-1}{}' = xs^{-1}x^{-1} = (xsx^{-1})^{-1} = (s')^{-1}$.

If x is in S, then S' is automorphic with S. Finally, if S' is automorphic with S for every x in G, then S' is called an invariant subgroup or normal divisor of G.

Now starting with an invariant subgroup S, we can define multiplication uniquely in the factored set G=S+aS+ .... . S and its cosets form a group called the quotient (or factor) group with S as the unit element. (aS)S = a(SS) = aS. (aS)(bS) = cS if ab = c.

Example:

Consider the 6-element permutation group on page 7 . The list of subgroups apart from the trivial ones $\{e\}$ and $\{G\}$ is

$$S_1 = \{e, \lambda\}, \quad S_2 = \{e, \mu\}, \quad S_3 = \{e, \nu\}, \quad S_4 = \{e, \alpha, \beta\}$$

To test whether S_n is an invariant subgroup, we form left and right cosets. If they are the same complexes, then S_n is invariant.

Subgroup	Left Cosets	Right Cosets
$S_1 \left\{ {e \atop \lambda} \right\}$	$\alpha S_1 \left\{ {\alpha \atop \nu} \right\}$, $\beta S_1 \left\{ {\beta \atop \mu} \right\}$	$S_1 \alpha \left\{ {\alpha \atop \mu} \right\}$ $S_2 \beta \left\{ {\beta \atop \nu} \right\}$
$S_4 \left\{ {e \atop {\alpha \atop \beta}} \right\}$	$\lambda S_4 \left\{ {\lambda \atop {\mu \atop \nu}} \right\}$	$S_4 \lambda \left\{ {\lambda \atop {\nu \atop \mu}} \right\}$

S_4 is an invariant subgroup. Thus we write $G = S_4 + \lambda S_4$. The group table for the quotient group is

	S_4	λS_4
S_4	S_4	λS_4
λS_4	λS_4	S_4

MAJOR EXAMPLE

To display many of the abstract definitions and theorems proved so far, we consider the following 12-element group which may be arrived at by considering all of the symmetry operations, excluding reflections, which leave a tetrahedron invariant. These consist of rotations of $\pm 120^\circ$ about each of the four three-fold axes which pass through a vertex and the center of the face opposite, rotations of 180° about each of the three two-fold axes through the centers of two opposite edges, and finally the identity operation. This group is isomorphic with a group which consists of some of the permutations of four objects. This is easily seen by labeling the four

corners of the tetrahedron as in the figure and observing
the changes which occur in the labeling as each symmetry
operation is applied. These are

Three-fold rotations:

$$a_1 = \begin{pmatrix} 1 & 2 & 3 & 4 \\ 1 & 3 & 4 & 2 \end{pmatrix}, \quad b_1 = \begin{pmatrix} 1 & 2 & 3 & 4 \\ 1 & 4 & 2 & 3 \end{pmatrix}, \quad a_2 = \begin{pmatrix} 1 & 2 & 3 & 4 \\ 4 & 2 & 1 & 3 \end{pmatrix}, \quad b_2 = \begin{pmatrix} 1 & 2 & 3 & 4 \\ 3 & 2 & 4 & 1 \end{pmatrix},$$

$$a_3 = \begin{pmatrix} 1 & 2 & 3 & 4 \\ 2 & 4 & 3 & 1 \end{pmatrix}, \quad b_3 = \begin{pmatrix} 1 & 2 & 3 & 4 \\ 4 & 1 & 3 & 2 \end{pmatrix}, \quad a_4 = \begin{pmatrix} 1 & 2 & 3 & 4 \\ 3 & 1 & 2 & 4 \end{pmatrix}, \quad b_4 = \begin{pmatrix} 1 & 2 & 3 & 4 \\ 2 & 3 & 1 & 4 \end{pmatrix}.$$

Two-fold rotations:

$$c_1 = \begin{pmatrix} 1 & 2 & 3 & 4 \\ 2 & 1 & 4 & 3 \end{pmatrix}, \quad c_2 = \begin{pmatrix} 1 & 2 & 3 & 4 \\ 3 & 4 & 1 & 2 \end{pmatrix}, \quad c_3 = \begin{pmatrix} 1 & 2 & 3 & 4 \\ 4 & 3 & 2 & 1 \end{pmatrix}$$

Identity:

$$e = \begin{pmatrix} 1 & 2 & 3 & 4 \\ 1 & 2 & 3 & 4 \end{pmatrix}$$

There are, of course, $4! = 24$ permutations of four objects.
To consider only half of these by excluding for example
$\begin{pmatrix} 1 & 2 & 3 & 4 \\ 1 & 2 & 4 & 3 \end{pmatrix}$ is equivalent to excluding the corresponding reflections. To understand that this exclusion does in fact
cut the number of elements in half it is sufficient simply
to notice that we might as easily have chosen our original
labeling of the corners to be exactly what we have, but
with 4 and 3 exchanged. Now since the second labeling
cannot be reached from the first by any rotations considered, and since the two labelings are equally acceptable,
each separate group must contain exactly half of the elements of the larger group of rotations and reflections.
(This is because any other labeling is obtained from one
or the other of these two and the group of rotations).
Here we have an example of the generation of a larger group
by the inclusion of a single new element, in this case
$\begin{pmatrix} 1 & 2 & 3 & 4 \\ 1 & 2 & 4 & 3 \end{pmatrix}$ and all its products with the original elements.
This can be viewed as the product of one group with another,
since $\begin{pmatrix} 1 & 2 & 3 & 4 \\ 1 & 2 & 4 & 3 \end{pmatrix}$ and the identity do indeed form a group.

Group Table

left \ right	e	a_1	b_1	a_2	b_2	a_3	b_3	a_4	b_4	c_1	c_2	c_3
e	e	a_1	b_1	a_2	b_2	a_3	b_3	a_4	b_4	c_1	c_2	c_3
a_1	a_1	b_1	e	b_3	c_2	b_4	c_3	b_2	c_1	a_3	a_4	a_2
b_1	b_1	e	a_1	c_3	a_4	c_1	a_2	c_2	a_3	b_4	b_2	b_3
a_2	a_2	b_4	c_2	b_2	e	b_1	c_1	b_3	c_3	a_4	a_3	a_1
b_2	b_2	c_3	a_3	e	a_2	c_2	a_4	c_1	a_1	b_3	b_1	b_4
a_3	a_3	b_2	c_3	b_4	c_1	b_3	e	b_1	c_2	a_1	a_2	a_4
b_3	b_3	c_1	a_4	c_2	a_1	e	a_3	c_3	a_2	b_2	b_4	b_1
a_4	a_4	b_3	c_1	b_1	c_3	b_2	c_2	b_4	e	a_2	a_1	a_3
b_4	b_4	c_2	a_2	c_1	a_3	c_3	a_1	e	a_4	b_1	b_3	b_2
c_1	c_1	a_4	b_3	a_3	b_4	a_2	b_1	a_1	b_2	e	c_3	c_2
c_2	c_2	a_2	b_4	a_1	b_3	a_4	b_2	a_3	b_1	c_3	e	c_1
c_3	c_3	a_3	b_2	a_4	b_1	a_1	b_4	a_2	b_3	c_1	c_2	e

Classes

To find other members of the class of a_1 for example, we form $x a_1 x^{-1} = q$ and continue with all new q. Evidently we need not use for x any element which commutes with a_1. For example:

$$a_2 a_1 a_2^{-1} = a_2 a_1 b_2 = a_2 c_2 = a_3$$
$$b_2 a_1 b_2^{-1} = b_2 a_1 a_2 = b_2 b_3 = a_4, \text{ etc.}$$

It turns out that the class of a_1 is $\{a_1, a_2, a_3, a_4\}$. Since $b_i = a_i^{-1}$, it follows by taking the inverse of each of the above equations that the class of b_1 is $\{b_1, b_2, b_3, b_4\}$. $(a_2 a_1 a_2^{-1})^{-1} = (a_3)^{-1} = a_2 a_1^{-1} a_2^{-1} = a_3^{-1}$ or $a_2 b_1 a_2^{-1} = b_3$. The classes then are $\mathbf{C}_e = \{e\}$, $\mathbf{C}_a = \{a_1, a_2, a_3, a_4\}$, $\mathbf{C}_b = \{b_1, b_2, b_3, b_4\}$, $\mathbf{C}_c = \{c_1, c_2, c_3\}$.

Subgroups

Since $12 = 1.2^2.3$, these must be of order, 1, 2, 3, 4, 6, 12.

$$S_1 = \{e\}$$
$$S_2 = \{e,c_1\},\ \{e,c_2\},\ \{e,c_3\}$$
$$S_3 = \{e,a_1,b_1\},\ \{e,a_2,b_2\},\ \{e,a_3,b_3\},\ \{e,a_4,b_4\}$$
$$S_4 = \{e,c_1,c_2,c_3\}$$
$$S_6 = \text{none}$$
$$S_{12} = \{G\}.$$

Invariant Subgroups

We form left and right cosets with a given element x, not in S_i. If they are different, then S_i is not invariant. Thus S_2 is not invariant, since

$$a_1 \cdot \begin{Bmatrix} e \\ c_1 \end{Bmatrix} = \begin{Bmatrix} a_1 \\ a_3 \end{Bmatrix} \text{ while } \begin{Bmatrix} e \\ c_1 \end{Bmatrix} a_1 = \begin{Bmatrix} a_1 \\ a_2 \end{Bmatrix}$$

The invariant subgroups are S_1, S_4, and S_{12}. In fact, it is easy to prove that if an invariant subgroup contains one element of a given class, then it must contain every element of that class. Thus S_2 and S_3 are excluded. Similarly it can be shown that any subgroup made up of complete classes is an invariant subgroup.

Factored Group of S_4

We multiply S_4 by an element not in S_4, for example, a_1. This would give $\{a_1,a_2,a_3,a_4\}$, a complex we call A. Next we take an element not in S_4 or A, for example b_1, and get the complex B = $\{b_1,b_2,b_3,b_4\}$. This exhausts the group and we may write

$$G = S_4 + A + B.$$

If we write $S_4 = E$, then $\{E,A,B\}$ forms a group under the rule for multiplying complexes together, which is isomorphic with S_3 or Z_3.

	E	A	B
E	E	A	B
A	A	B	E
B	B	E	A

This group is homomorphic with G itself if we associate the elements of S_4 with the element E, the elements of A with the element A, and the elements of B with the element **B**.

Multiplication of Classes

This operation is similar to that of multiplying complexes except that we do not discard elements which duplicate others. Thus $C_a C_a = \{a_1, a_2, a_3, a_4\}\{a_1, a_2, a_3, a_4\} = \{4b_1, 4b_2, 4b_3, 4b_4\} = 4C_b$. The multiplication table for the classes of G is

	C_e	C_a	C_b	C_c	sum in row
C_e	C_e	C_a	C_b	C_c	G
C_a	C_a	$4C_b$	$4C_e + 4C_c$	$3C_a$	$4G$
C_b	C_b	$4C_e + 4C_c$	$4C_a$	$3C_b$	$4G$
C_c	C_c	$3C_a$	$3C_b$	$3C_e + 2C_c$	$3G$

Multiplication of classes is always commutative.

Proof: From the definition of class it is obvious that

$$a C_b a^{-1} = C_b$$

that is multiplication of a class by a on the right and by a^{-1} on the left reproduces the class. Then

$$a C_b = C_b a$$

and summing over all elements in the class C_a to which a

belongs,

$$c_a c_b = c_b c_a.$$

Obviously classes do not form a group under class multiplication. A check on the class multiplication table can be made easily in the following way. If there are n elements in a certain class C, and if we multiply C by all the other classes, that is with G, we must reproduce G n times. So if we add up the rows or columns, each class must appear in the sum n times.

II

Theory of Representations

<u>Definition</u>:

(a) A set of matrices under matrix multiplication, $\{A_1 (a_1), A_2 (a_2) \ldots A_n (a_n)\}$, which is homomorphic with the group $\{a_1, a_2 \ldots a_n\}$ under its operation, is said to be a <u>representation</u> of the group. The <u>order</u> of the matrices is called the <u>degree</u> or <u>dimension</u> of the representation. Note that matrix representations are not the only representations of abstract groups. Thus matrices of order 3 with determinant +1 form a representation of the proper orthogonal group $[O^+(3)]$, while forms of degree n in x_1, x_2, x_3 also form a representation of $O^+(3)$ called a polynomial representation. In this book, however, we are almost exclusively concerned with matrix representations. Evidently the matrices must be square and non-singular to satisfy the group postulates. As a trivial example, the unit matrix of any order may be associated homomorphically with every element of any group, and so it is a representation of that group. That is, if we write $A_1 = I$, $A_2 = I$, etc. and $a_1 a_2 = a_m$ then $A_1 A_2 = A_m$

(b) There is no unique representation of a group. Thus if $\{A_i\}$ and $\{B_i\}$ are both representations of dimension n and m, of G, respectively, then the set of matrices $\left\{\begin{pmatrix} A_i & 0 \\ 0 & B_i \end{pmatrix}\right\}$ is also a representation of G of dimension n + m. The zeros represent n x m and m x n null matrices.

(c) Furthermore, if $\{A_i\}$ is an nth degree representation of G, then so is $\{XA_iX^{-1}\}$ where X is a non-singular n x n matrix. We say that $\{B_i\}$ and $\{A_i\}$ are _equivalent_ representations if there exists some matrix $\bar{X}$ such that $B_i = \bar{X} A_i \bar{X}^{-1}$ for all i.

(d) A representation is said to be _faithful_ if the application is an isomorphism.

(e) Reducibility. A representation is said to be _reducible_ if an equivalent representation exists in which each matrix M_i has the form

$$M_i = \begin{pmatrix} A_i & C_i \\ 0 & B_i \end{pmatrix}$$

where A_i is an n x n matrix, B_i is an m x m matrix, C_i is a rectangular n x m matrix, and the zero represents an m x n null matrix. When such an equivalent representation has been found, we say that the original representation M has been reduced to the representations A and B, because

$$M_iM_j = M_k \Rightarrow \begin{cases} A_iA_j = A_k \\ B_iB_j = B_k. \end{cases}$$

A and B may themselves be further reducible. If no such equivalent representation exists, the original one M is said to be an _irreducible representation_ of G. A represent-ation is said to be _fully reducible_ if an equivalent repre-sentation exists in which each matrix M_i' has the form

$$M_i' = \begin{pmatrix} A_i & 0 \\ 0 & B_i \end{pmatrix}$$

that is, it is reducible and in addition $C_i = 0$. For unitary representations there is no distinction; that is, all reducible unitary representations are fully reducible.

IMPORTANT THEOREMS

The following theorems will be proved <u>only for finite groups</u> of order h. They can, however, be extended to some infinite groups - the so-called <u>compact</u> groups.

<u>Theorem I</u>: Any representation of G is equivalent, by means of a similarity transformation, to a unitary representation. (A unitary representation is one such that for all i, $A_iA_i^\dagger = A_i^\dagger A_i = I$.)

<u>Proof</u>: Let $\{A_i\}$ be a representation of G. Construct the Hermitian matrix $H = \sum_{i=1}^{h} A_iA_i^\dagger$

$$H^\dagger = [\sum_{i=1}^{h}(A_iA_i^\dagger)]^\dagger = \sum_{i=1}^{h}(A_iA_i^\dagger)^\dagger = \sum_{i=1}^{h}(A_i^\dagger)^\dagger A_i^\dagger = \sum_{i=1}^{h} A_iA_i^\dagger = H.$$

According to matrix algebra, any Hermitian matrix H may be diagonalized by a similarity transformation whose matrix U is a unitary matrix formed in columns by the orthonormal eigenvectors of H. Let

$$d = U^{-1}HU = \sum_i U^{-1}A_iA_i^\dagger U = \sum_i U^{-1}A_iUU^{-1}A_i^\dagger U.$$

Hence $d = \sum_i (U^{-1}A_iU)(U^\dagger A_iU^{-1\dagger})^\dagger = \sum_i (U^{-1}A_iU)(U^{-1}A_iU)^\dagger$

$$d = \sum_i A_i'A_i'^\dagger, \text{ where } A_i' = U^{-1}A_iU.$$

Now d is diagonal by definition and its diagonal elements are real and positive.

$$d_{jj} = \sum_{i,k} a'^{(i)}_{jk} (a'^{(i)\dagger})_{kj} = \sum_{i,k} a'^{(i)}_{jk} a'^{(i)*}_{jk} = \sum_{i,k} |a'^{(i)}_{jk}|^2 \geq 0$$

In fact, it can never equal zero, since if $a'^{(i)}_{jk} = 0$ for all k, A'_i would be singular. Since d is diagonal and so behaves like a number, we may define <u>non-linear functions</u> of d. In particular we define $d^{\frac{1}{2}}$ and $d^{-\frac{1}{2}}$ by taking the respective powers of the diagonal elements. Evidently $(d^{\frac{1}{2}})^2 = d$ and $(d^{\frac{1}{2}})(d^{-\frac{1}{2}}) = I$, where I is the unit matrix. Diagonal matrices commute with one another, and we can write

$$d^{-\frac{1}{2}} d d^{-\frac{1}{2}} = d^{-\frac{1}{2}} \sum_{i=1}^{h} A'_i A'^{\dagger}_i d^{-\frac{1}{2}}$$

$$I = d^{-\frac{1}{2}} \sum_{i}^{h} A'_i A'^{\dagger}_i d^{-\frac{1}{2}}.$$

By the rearrangement theorem, $\{A'_j A'_i\}_{\text{all i, one j}}$ is equal to $\{A'_i\}_{\text{all i}}$. Hence

$$I = d^{-\frac{1}{2}} \sum_{i=1}^{h} (A'_j A'_i)(A'_j A'_i)^{\dagger} d^{-\frac{1}{2}}$$

$$= d^{-\frac{1}{2}} \sum_{i=1}^{h} (A'_j A'_i A'^{\dagger}_i A'^{\dagger}_j) d^{-\frac{1}{2}}$$

$$= d^{-\frac{1}{2}} A'_j d^{\frac{1}{2}} d^{-\frac{1}{2}} \underbrace{\sum_{i=1}^{h} A'_i A'^{\dagger}_i d^{-\frac{1}{2}}}_{= d} d^{\frac{1}{2}} A'^{\dagger}_j d^{-\frac{1}{2}}$$

$$= I$$

$$I = (d^{-\frac{1}{2}} A'_j d^{\frac{1}{2}})(d^{-\frac{1}{2}} A'_j d^{\frac{1}{2}})^{\dagger}$$

Finally, $\qquad I = (d^{-\frac{1}{2}} U^{-1} A_j U d^{\frac{1}{2}})(d^{-\frac{1}{2}} U^{-1} A_j U d^{\frac{1}{2}})^{\dagger}$

We have thus found a transformation matrix $U' = Ud^{\frac{1}{2}}$ which transforms $\{A_i\}$ into the equivalent unitary set $\{A_i'\} = \{U'^{-1}A_iU'\}$.

Note: From now on we shall be concerned only with unitary representations, and the word representation will be used to mean a unitary representation.

Theorem II: Schur's Lemma

If there exists a matrix M, not necessarily one of the representation matrices $\{A_i\}$, such that $MA_i = A_iM$ for all i, then

        a. if $\{A_i\}$ is irreducible, M is a constant matrix; that is, $M = cI$ where c is a number.

        b. if M is not a constant matrix, then $\{A_i\}$ is reducible. (This is merely the contra-positive of a.)

Proof: Let there be a matrix M such that for all i

(i) $A_iM = MA_i$, where the A_i are unitary matrices. Then, we get, by taking the hermitian conjugate,

$$M^\dagger A_i^{\ \dagger} = A_i^{\ \dagger}M^\dagger$$

$$A_i(M^\dagger A_i^{\ \dagger})A_i = A_i(A_i^{\ \dagger}M^\dagger)A_i$$

(ii) $A_iM^\dagger = M^\dagger A_i$.

By adding and subtracting (i) and (ii), we get

$$A_i(M+M^\dagger) = (M+M^\dagger)A_i; \ M+M^\dagger = H_1 \left.\begin{array}{c} \\ \\ \end{array}\right\} \begin{array}{l} \text{are} \\ \text{hermitian.} \end{array}$$
$$A_i i(M-M^\dagger) = i(M-M^\dagger)A_i; \ i(M-M^\dagger) = H_2$$

        Thus far we have proved that if M exists and commutes with $\{A_i\}$, then two hermitian matrices, H_1 and H_2, exist and commute with $\{A_i\}$. Now if we can prove that the existence of a commuting hermitian matrix H implies the consequences of this theorem, we have indeed proved the theorem for all M.

Since H is hermitian, it can be diagonalized by some U, $d = U^{-1}HU$. Let $A_i' = U^{-1}A_i U$.

$$A_i H = HA_i$$

$$U^{-1}A_i UU^{-1}HU = U^{-1}HUU^{-1}A_i U$$

$$A_i' d = dA_i'$$

What we now must show is that the diagonal elements of d are all identical if $\{A_i\}$ is irreducible. We look at the $\mu\nu$ element of the matrix product.

$$\sum_\lambda a'^{(i)}_{\mu\lambda} \delta_{\lambda\nu} d_\nu = \sum_\varkappa d_\mu \delta_{\mu\varkappa} a'^{(i)}_{\varkappa\nu}$$

$$a'^{(i)}_{\mu\nu} d_\nu = d_\mu a'^{(i)}_{\mu\nu}$$

(iii) $$a'^{(i)}_{\mu\nu}(d_\nu - d_\mu) = 0, \text{ all } i, \mu, \nu.$$

The conclusion drawn at this point in text books is that, since $\{A_i\}$ is irreducible, $a'^{(i)}_{\mu\nu} \neq 0$ for some i and for every μ and ν. Hence $d_\nu = d_\mu$ for all μ and ν. We can see that we need to state this more clearly in the following example. If in a 3 x 3 representation the (1,2) element of every matrix A_i is zero, the matrices are not yet in block form. Yet by looking at the (1,2) element alone, we would conclude that it is possible to have $d_1 \neq d_2$, even though the representation $\{A_i\}$ appears irreducible.

Actually the proof does follow from equation (iii), but we must be more careful. We follow it out for the 3 x 3 case first and then do it in general. Suppose $d_1 \neq d_2$. Then for all i, not only must $a'^{(i)}_{12} = 0$, but at least one of the pair $a'^{(i)}_{13}$, $a'^{(i)}_{23}$ must also equal zero. If not, then $d_1 = d_3$ and $d_2 = d_3$; hence $d_1 = d_2$. But if one of this pair is also zero for all i, then the matrices either are or may easily be transformed into block form, because we have at least two zero elements above the diagonal.

X	0	0
0	X	X
0	X	X

The general proof follows the same argument. Suppose we have an ℓ-dimensional representation and suppose for some μ, ν, $d_\mu \neq d_\nu$. Then by equation (iii), $a'^{(i)}_{\mu\nu} = 0$ for all i. But also because of the transitive property of equality, at least one of every pair $a'^{(i)}_{\mu\lambda}$, $a'^{(i)}_{\nu\lambda}$ must also be zero for all i. There are ℓ-2 such pairs since we do not count $\lambda = \mu$, $\lambda = \nu$ giving a total of $\ell - 1$ zeros at least. Then by a similarity transformation, we bring $\{A'_i\}$ into block form.

$$\begin{array}{c|cccc} & 0 & 0 & \dots & 0 \\ \hline 0 & & & & \\ 0 & & & & \\ \vdots & & & & \\ 0 & & & & \end{array}$$

So if M is not a constant matrix, then $\{A_i\}$ is reducible.

Theorem III:

If we are given two irreducible representations of G, $\{A_i\}$ of dimension ℓ_a and $\{B_i\}$ of dimension ℓ_b, and a rectangular $\ell_a \times \ell_b$ matrix M such that $A_i M = M B_i$ for all i, then

(a) if $\ell_a \neq \ell_b$, M = 0

or (b) if $\ell_a = \ell_b$ either M = 0 or the determinant of M is not zero, $|M| \neq 0$.

Proof:

We restrict our proof to unitary representations by Theorem I.

$$A_i M = M B_i \text{ , all i, given.}$$
$$M^\dagger A_i^\dagger = B_i^\dagger M^\dagger \text{ hermitian conjugation}$$
$$M^\dagger A_i^{-1} = B_i^{-1} M^\dagger \text{ unitarity}$$

Multiplying on the left by M, we get

$$MM^\dagger A_i^{-1} = MB_i^{-1}M^\dagger$$

By the group properties B_i^{-1} is some B_j and A_i^{-1} is the corresponding A_j, and so by the postulate of this theorem, A_jM = MB_j; that is, $A_i^{-1}M = MB_i^{-1}$. Making this substitution in the right hand side of the last line given above, we get

$$MM^\dagger A_i^{-1} = A_i^{-1}MM^\dagger.$$

Thus $MM^\dagger$ is a matrix which satisfies Schur's Lemma, Theorem II. Since $\{A_i\}$ is irreducible, it follows that $MM^\dagger$ is a constant matrix; that is, $MM^\dagger = cI$.

Case (b): $\ell_a = \ell_b$. $|M|$ exists. We take determinants of both sides of the last equation:

$$|MM^\dagger| = |cI|; \quad |M||M^\dagger| = c^{\ell_a} \text{ or } \|M\|^2 = c^{\ell_a}[|M||M|^* = \|M\|^2]$$

Now either $c=0$ or $c \neq 0$. If $c \neq 0$, then $|M| \neq 0$. If $c= 0$, then $MM^\dagger = 0$, or

$$\sum_\lambda M_{\mu\lambda}(M^\dagger)_{\lambda\nu} = 0$$

$$\sum_\lambda M_{\mu\lambda}M_{\nu\lambda}^* = 0 \quad \text{from definition of hermitian conjugate}$$

$$\sum_\lambda |M_{\mu\lambda}|^2 = 0 \quad \text{diagonal element, } \mu = \nu$$

Since this is a positive definite expression, every term must vanish separately. Hence $M = 0$.

Case (a): $\ell_a \neq \ell_b$. Let $\ell_a < \ell_b$. The equations $A_iM = MB_i$ and $MM^\dagger = cI$ can be pictured as

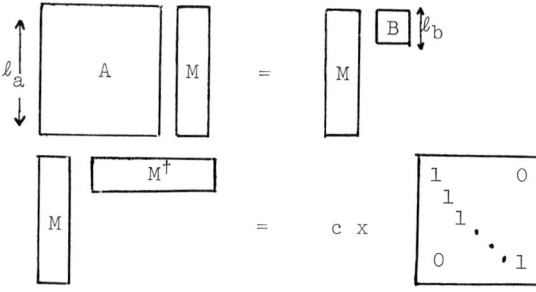

Now a square $\ell_a \times \ell_a$ matrix N can be defined so that its first ℓ_b columns are the ℓ_b columns of M and the rest are zeros

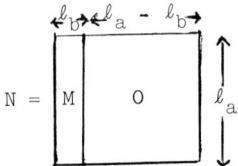

$$N = \begin{bmatrix} M & O \end{bmatrix}$$

Thus

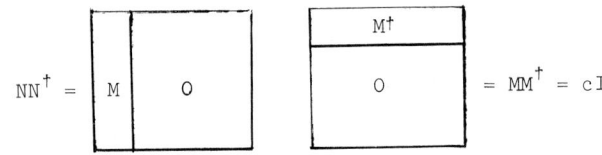

$$NN^\dagger = \begin{bmatrix} M & O \end{bmatrix} \begin{bmatrix} M^\dagger \\ O \end{bmatrix} = MM^\dagger = cI$$

N is a square matrix whose determinant vanishes since it contains at least one column of zeros.

$$NN^\dagger = cI$$
$$|NN^\dagger| = c^{\ell_a} \quad \text{taking determinant}$$
$$|N||N|^* = c^{\ell_a} \quad \text{separating determinant}$$
$$0 = c^{\ell_a} \quad \text{of product}$$
$$|N| = 0$$
$$0 = c$$

Hence $NN^\dagger = 0$; and as we saw in case (b), this implies $N = 0$. Since M is contained in N, $M = 0$. Q.E.D.

Corollary to case (b): If $\ell_a = \ell_b$ and $M \neq 0$, then $\{A_i\}$ and $\{B_i\}$ are equivalent to one another.

Proof: In theorem III we have shown that the condition $\ell_a = \ell_b$ leads to the consequence that if $M \neq 0$, then $|M| \neq 0$ (where M is such that $A_iM = MB_i$, all i). Since $|M| \neq 0$, M^{-1} exists. We multiply the equation of the postulate of III from the left by M^{-1}.

$$M^{-1}(A_iM) = M^{-1}(MB_i) = B_i \quad \text{for all i}$$

Thus $\{B_i\}$ is equivalent to $\{A_i\}$ and since $MM^\dagger = c^{\ell_a}I$, it follows that $c^{-\ell_a/2}$ M is a unitary matrix.

<u>Theorem IV:</u> <u>Orthogonality Theorem</u>

<u>Notation:</u> Let $\Gamma^{(i)}(R)_{\mu\nu}$ be the (μ,ν) matrix element of the ith irreducible unitary representation of the element R of G. The range of R will be the h elements of G. The range of i will be the number of inequivalent irreducible representations. The range of μ and ν will be the dimension ℓ_i of the ith irreducible representation. When we write $\Gamma^{(i)}(R)$, we mean the entire matrix corresponding to the element R in the ith representation.

<u>Theorem:</u> $\displaystyle\sum_R \Gamma^{(i)}(R)_{\mu\nu}^* \Gamma^{(j)}(R)_{\alpha\beta} = \frac{h}{\ell_i}\delta_{ij}\delta_{\mu\alpha}\delta_{\nu\beta}.$

<u>Proof:</u> <u>Case (a)</u> $i \neq j$. Let $\Gamma^{(1)}(R)$ not be equivalent to $\Gamma^{(2)}(R)$. First we define a specific matrix M, prove that it satisfies the postulates of Theorem III, and thus show that M = 0.

Let $\quad M = \displaystyle\sum_R \Gamma^{(2)}(R) \; X \; \Gamma^{(1)}(R^{-1}).$

where X is any rectangular or square matrix which makes the above matrix product possible.

Then $\Gamma^{(2)}(S)M = \displaystyle\sum_R \Gamma^{(2)}(S)[\Gamma^{(2)}(R)X\Gamma^{(1)}(R^{-1})]\Gamma^{(1)}(S^{-1})\Gamma^{(1)}(S)$

$\qquad = \displaystyle\sum_R [\Gamma^{(2)}(S)\Gamma^{(2)}(R)]X[\Gamma^{(1)}(R^{-1})\Gamma^{(1)}(S^{-1})]\Gamma^{(1)}(S)$

If SR = T, then $\Gamma^{(i)}(S)\Gamma^{(i)}(R)=\Gamma^{(i)}(SR)=\Gamma^{(i)}(T)$

$\qquad\qquad \Gamma^{(i)}(R^{-1})\Gamma^{(i)}(S^{-1})=\Gamma^{(i)}(R^{-1}S^{-1})=\Gamma^{(i)}(T^{-1})$

and by the rearrangement theorem, the sum over R is the same, for a particular S, as that over SR = T. Hence we get the result

$\qquad\qquad \Gamma^{(2)}(S)M = \displaystyle\sum_T \Gamma^{(2)}(T)X\Gamma^{(1)}(T^{-1})\Gamma^{(1)}(S)$

$\qquad\qquad\qquad = M\Gamma^{(1)}(S) \text{ for all S.}$

Hence by Theorem III and its corollary, M = 0. Now choose X to be a matrix every element of which is zero except one, which we shall call $x_{\nu\alpha}$, and $x_{\nu\alpha} = 1$.

$$M = 0 = \sum_R \Gamma^{(2)}(R) X \Gamma^{(1)}(R^{-1})$$

$$0 = \sum_{\gamma\delta} \sum_R \Gamma^{(2)}(R)_{\mu\gamma} X_{\gamma\delta} \Gamma^{(1)}(R^{-1})_{\delta\beta}$$

$$0 = \sum_R \Gamma^{(2)}(R)_{\mu\nu} \Gamma^{(1)}(R^{-1})_{\alpha\beta}$$

$$0 = \sum_R \Gamma^{(1)}(R)_{\beta\alpha}^* \Gamma^{(2)}(R)_{\mu\nu}$$

The last step follows from the unitarity of the representation.

$$\Gamma^{(i)}(R^{-1})_{\eta\xi} = [\Gamma^{(i)}(R)]_{\eta\xi}^{-1} = \Gamma^{(i)}(R)_{\xi\eta}^*$$

or $\quad \Gamma^{(i)}(R^{-1}) = [\Gamma^{(i)}(R)]^{-1} = \Gamma^{(i)}(R)^\dagger \equiv \widetilde{\Gamma^{(i)}(R)}^*.$

This proves the theorem for $i \neq j$.

Case (b): $i = j = 1$ We define $M = \sum_R \Gamma^{(1)}(R) X \Gamma^{(1)}(R^{-1})$.

Then just as in case (a), $M\Gamma^{(1)}(S) = \Gamma^{(1)}(\check{S})M$ for all S. Hence by Schur's lemma, Theorem III, $M = cI$. Again we let $x_{\nu\nu'} = 1$ for a particular ν and ν' and choose all other elements of X equal to zero.

$$M = cI = \sum_R \Gamma^{(1)}(R) X \Gamma^{(1)}(R^{-1})$$

$$(i) \qquad c\delta_{\mu\mu'} = \sum_R \Gamma^{(1)}(R)_{\mu\nu} \Gamma^{(1)}(R^{-1})_{\nu'\mu'}$$

Let $\mu = \mu'$ and sum on μ

$$c\sum_{\mu=1}^{\ell_1} 1 = \sum_R \sum_\mu \Gamma^{(1)}(R^{-1})_{\nu'\mu} \Gamma^{(1)}(R)_{\mu\nu}$$

$$c\ell_1 = \sum_R \Gamma^{(1)}(R^{-1}R)_{\nu'\nu} = \sum_R \Gamma^{(1)}(E)_{\nu'\nu}$$

$$= \sum_R \delta_{\nu'\nu} = h\delta_{\nu'\nu}$$

$$\therefore \quad c = \frac{h}{\ell_1}\delta_{\nu\nu'}$$

We replace c in (i) by this value and rewrite $\Gamma^{(1)}(R^{-1})_{\nu'\mu'}$ as $\Gamma^{(1)}(R)^*_{\mu'\nu'}$, as before to get

$$\sum_R \Gamma^{(1)}(R)^*_{\mu'\nu'}\Gamma^{(1)}(R)_{\mu\nu} = \frac{h}{\ell_1}\delta_{\mu\mu'}\delta_{\nu\nu'}$$

This proves the theorem for i = j.

This central theorem of representation theory has a useful geometrical interpretation. $\Gamma^{(i)}(R)_{\mu\nu}$ can be thought of as the Rth component in an h-dimensional space of a vector which is distinguished from other vectors of different i, μ, and ν. The theorem then states that all such <u>distinct</u> vectors are orthogonal in this space. How many such distinct vectors are there? For a given i there are ℓ_i^2 pairs $\mu\nu$. Hence the total number of distinct vectors is $\sum \ell_i^2$, where the sum is over all inequivalent irreducible representations. But there can be no more than h distinct orthogonal vectors in an h-dimensional space. Hence $\sum \ell_i^2 \leq h$. Later we shall prove in fact that the equality holds exactly. This is a very useful fact in working out representations of any group. For a group of order 6, for example, the set $\{\ell_i\}$ <u>must</u> be $\{1,1,2\}$, because no other combination of integers satisfies this equation. Hence if we find two one-dimensional and one two-dimensional irreducible representations, we have found all the irreducible representations of that group.

The Characters of a Representation

We need simple criteria for distinguishing one

irreducible representation from another and for decomposing
reducible representations into their irreducible components.
The actual appearance of two equivalent irreducible representations may be quite different from one another and yet
not especially distinguishable at a glance from a third
reducible representation. Our first problem is then to be
able to distinguish at a glance irreducible representations
which are distinct from one another, (i.e., not equivalent)
and also to be able to see whether a given representation
is reducible or irreducible. The place to look for these
criteria is the theory of the invariants of non-singular
matrices. For a diagonalizable matrix for example, the
eigenvalues are a set of invariants that completely char-
acterize the matrix. To prove that they are indeed invariant under similarity transformations, we write the secular
equation and transform it. Let A be the matrix in question
and let $A' = XAX^{-1}$.

$$(i) \qquad\qquad |\, A - \lambda I \,| = 0$$

$$|\, XX^{-1}|\ \ |\, A - \lambda I \,| = 0$$

$$|\, X \,|\,|\, A - \lambda I \,|\,|\, X^{-1} \,| = 0$$

$$|\, X(A - \lambda I)X^{-1}| = 0$$

$$|\, XAX^{-1} - X^{-1}\lambda IX \,| = 0$$

$$(ii) \qquad\qquad |\, A' - \lambda I \,| = 0$$

Thus any λ which satisfies (i) satisfies (ii) also. The
eigenvalues would therefore be a good choice since they
completely define the matrix. However, they are tedious
to determine.

　　　A closer look at (i) suggests another complete list
of invariants of a matrix namely, the coefficients of each
power of λ in the secular equation. For if the set $\{\lambda_i\}$
are invariants, so also must be the coefficients in the
polynomial equation whose roots are $\{\lambda_i\}$. For example, it
is easy to see that the coefficient of λ^0 is the determin-
ant $|\, A \,|$, which is indeed invariant under a similarity trans-

formation. But we must be careful here. The secular equation for an nth order matrix gives us $(n + 1)$ coefficients of $(\lambda^0, \lambda^1, \lambda^2, \lambda^3, \ldots \lambda^n)$. However the coefficient of λ^n is always 1, and that still leaves n quantities with which to characterize the matrix. Nevertheless, we may fix at will a value of $|A_1|$ without in any way altering the essential differences between the matrices of the set. This is the usual normalization argument. Since we have agreed upon unitary representations, we have already insisted that $\|A_1\| = 1$ for all i.

The simplest invariant coefficient of the secular equation is then the coefficient of λ^{n-1}, which is the trace of the matrix A, $TrA = \sum_\mu A_{\mu\mu}$. This gives enough useful information for virtually every application of group theory to physical problems.

Definition: The characters $\chi^{(i)}(R)$ of a representation i are the traces $\sum_\mu \Gamma^{(i)}(R)_{\mu\mu}$ of the matrices of the representation.

Theorem V: If R and S belong to the same class, they have identical characters.

Proof: Some element T exists such that $S = TRT^{-1}$. Consequently, $\Gamma^{(i)}(S) = \Gamma^{(i)}(T) \; \Gamma^{(i)}(R) \; \Gamma^{(i)}(T^{-1})$

$\chi^{(i)}(S) = Tr[\Gamma^{(i)}(T)\Gamma^{(i)}(R)\Gamma^{(i)}(T^{-1})] = Tr[\Gamma^{(i)}(T^{-1})\Gamma^{(i)}(T)\Gamma^{(i)}(R)]$

$= Tr[\Gamma^{(i)}(T^{-1}T)\Gamma^{(i)}(R)] = Tr[I\Gamma^{(i)}(R)]$

$= Tr[\Gamma^{(i)}(R)] = \chi^{(i)}(R)$

A similar proof holds for a more general theorem which says that R and S have identical matrix invariants.

Definition: $\chi^{(i)}(C_k)$ is a number which is equal to the character of any element in the class C_k.

Theorem VI: First Orthogonality Relation If i and j are irreducible and N_k is the number of elements in C_k, then

$$\sum_k \chi^{(i)}(C_k)^* \chi^{(j)}(C_k) N_k = h\delta_{ij}$$

Proof: Starting from the orthogonality relations, Theorem IV, and setting $\mu = \nu$, $\beta = \alpha$, and summing on μ and α, we get

$$\sum_R \chi^{(i)*}(R)\chi^{(j)}(R) - \frac{h}{\ell_i}\delta_{ij}\underbrace{\sum_{\mu,\alpha}\delta_{\mu\alpha}\delta_{\mu\alpha}}_{=\ell_i}$$

$$\sum_k \chi^{(i)}(\mathbf{C}_k)^*\chi^{(j)}(\mathbf{C}_k)N_k = h\delta_{ij}. \qquad \text{Q.E.D.}$$

The last step involves regrouping the upper sum and applying Theorem V. This result can once again be interpreted as an orthogonality relation between all distinct vectors in a space of dimension equal to the number of classes in G. But now the vectors are distinguished from one another only by the label i. By the same geometrical argument as before, we conclude that the number of distinct irreducible representations must be less than or equal to the number of classes of G. In fact, it can be shown as a consequence of Theorem VII that

number of irreducible representations=number of classes.

Theorem VII: If $\Gamma^{(i)}$ is irreducible, then
$$\sum_i \chi^{(i)}(\mathbf{C}_k)^*\chi^{(i)}(\mathbf{C}_\ell) = \frac{h}{N_\ell}\delta_{k\ell}$$

This is the second orthogonality relation.

Proof: Construct matrices $Q_{ik} \equiv \chi^{(i)}(\mathbf{C}_k)^*$ and $Q'_{\ell j} \equiv \chi^{(j)}(\mathbf{C}_\ell)\frac{N_\ell}{h}$. Then by Theorem VI,

$$(QQ')_{ij} = \sum_k \chi^{(i)}(\mathbf{C}_k)^*\chi^{(j)}(\mathbf{C}_k)\frac{N_k}{h} = \delta_{ij}$$

by the first orthogonality relation. This is $QQ' = I$. If we assume Q and Q' are non-singular, then $Q' = Q^{-1}$. Hence $Q'Q = I$. That is, $\sum_i \chi^{(i)}(\mathbf{C}_k)^*\frac{N_k}{h}\chi^{(i)}(\mathbf{C}_\ell) = \delta_{k\ell}$ and the theorem follows.

This proof of Theorem VII is defective unless we also prove
that Q is not singular. Instead of doing that, we give a
different proof devoid of this defect. We recall that
n orthogonal vectors, $\chi^{(i)}$, completely span an n-dimensional
space in the sense that any vector χ in the space can be
written as

$$\chi = \sum_i a_i \chi^{(i)}$$

where $\{a_i\}$ are constant numbers. We define the weighted
inner product as

$$\chi \cdot \lambda = \sum_k \chi(c_k) \lambda(c_k)^* N_k$$

and we take the inner product of χ and $\chi^{(j)}$

$$\chi \cdot \chi^{(j)} = \sum_i a_i \chi^{(i)} \cdot \chi^{(j)}$$

$$= \sum_i a_i h \delta_{ij} = h a_j \quad \text{by the first ortho-} \atop \text{gonality relation.}$$

Hence, $\qquad \chi = \sum_i \frac{\chi \cdot \chi^{(i)}}{h} \chi^{(i)}$

$$= \chi \cdot \sum_i \frac{\chi^{(i)} \chi^{(i)}}{h}$$

Since this equation must be true for arbitrary χ, the
matrix on the right must be the unit matrix, properly defined for the weighted inner product. Let us write out
the last equation in component notation.

$$\chi(c_\ell) = \sum_{k,i} N_k \chi(c_k) \chi^{(i)}(c_k)^* \chi^{(i)}(c_\ell) \frac{1}{h}$$

$$0 = \sum_k \chi(c_k) \left[\frac{N_k}{h} \sum_i \chi^{(i)}(c_k^*) \chi^{(i)}(c_\ell) - \delta_{k\ell} \right]$$

Thus for arbitrary χ, the term in brackets vanishes for
each component k. This proves the second orthogonality

relation. In the sense of this proof, it can be called the Completeness Relation.

This proof evidently establishes the fact that $Q' = Q^{-1}$. And since Q^{-1} exists, $|Q| \neq 0$. Thus the <u>rows of Q are linearly independent and so are the columns</u>-a very important result. This establishes the fact that the characters of a given irreducible representation are unique and distinct from those of any other. Similarly the characters of a given class in different irreducible representations are unique and distinct from those of any other class. We see more of this in the following discussion of the character table.

THE CHARACTER TABLE

A character table has the form shown below

	$\mathbf{C}_1 = \mathbf{C}_E$	$N_2\mathbf{C}_2$	$N_3\mathbf{C}_3$	...
$\Gamma^{(1)}$	1	1	1	
$\Gamma^{(2)}$	ℓ_2	...	...	
$\Gamma^{(3)}$	ℓ_3	...	...	

It is merely the matrix Q defined in Theorem **VII**. It has all of the following properties and can usually be constructed for any group simply from a knowledge of the set $\{N_k\}$ and h.

(1) It is a square table, i.e., number of classes = number of irreducible representations.

(2) The first column contains the dimensions of each representation $\chi^{(i)}(\mathbf{C}_1) = \chi^{(i)}(E) = \mathrm{Tr}\Gamma^{(i)}(E) = \ell_i$.

(3) The sum of the squares of the terms in the first column is h.$\therefore \sum \ell_i^2 = h$. A proof of the equality follows from putting $k=\ell=1$ in the second orthogonality relation. A second proof will be given later (Theorem **IX**).

(4) Since $\Gamma^{(1)}(R) = 1$ for all R, $\chi^{(1)}(\mathbf{C}_k) = 1$ for all k. This is called the <u>unit</u> or <u>identity</u> representa-

tion.

      (5) Rows are orthogonal with inclusion of N_k as the weighting factor.

      (6) Columns are orthogonal.

      (7) If we know the <u>coefficients of class multi-plication</u>, c_{ijk} defined by the equation $\mathbf{C}_i\mathbf{C}_j = \sum_k c_{ijk}\mathbf{C}_k$, then it can be proved

$$N_j\chi^{(i)}(\mathbf{C}_j)N_k\chi^{(i)}(\mathbf{C}_k) = \ell_i\sum_\ell c_{jk\ell}N_\ell\chi^{(i)}(\mathbf{C}_\ell)$$

[proof omitted].

      (8) If we know the character table for the factor group FG, then we get some irreducible representations of G by giving to each class $\mathbf{C}_k$ of G the character of the set containing $\mathbf{C}_k$ in FG.

   <u>Examples</u>:

      (a) Character table for the three-fold permutation [isomorphic with point groups D_3 or 32, C_{3v} or $3m$]. There are three classes $\mathbf{C}_1 = e$; $\mathbf{C}_2 = \alpha, \beta$; and $\mathbf{C}_3 = \lambda, \mu, \nu$. The table must be 3 x 3. The first row contains the identity representation, and the first column is $1, \ell_2, \ell_3$, where

$$1 + \ell_1^2 + \ell_3^2 = 6.$$

Hence $\ell_2 = 1$, $\ell_3 = 2$.

	$\mathbf{C}_1$	$2\mathbf{C}_2$	$3\mathbf{C}_3$
Γ^1	1	1	1
Γ^2	1	a	b
Γ^3	2	c	d

We have four numbers to determine: a, b, c, d. We note the Γ^2 is one-dimensional. We recall that $S = \{e,\alpha,\beta\} = \mathbf{C}_1 + \mathbf{C}_2$ is an invariant subgroup of index 2. The factor group has multiplication table

	S	A
S	S	A
A	A	S

where $A = \{\lambda,\mu,\nu\} = \mathbf{C}_3$. The irreducible representations of the factor group must be $\{1,1\}$ and $\{1,-1\}$. Then from (8) above, we obtain Γ_2 by choosing $a = 1$ and $b = -1$. The first two rows satisfy the orthogonality relations. The orthogonality relation for columns may be used to determine c and d.

$$1 \times 1 + 1 \times a + 2.c = 0 \qquad \therefore \quad c = -1$$
$$1 \times 1 + 1 \times b + 2.d = 0 \qquad \therefore \quad d = 0$$

	$\mathbf{C}_1$	$2\mathbf{C}_2$	$3\mathbf{C}_3$
Γ^1	1	1	1
Γ^2	1	1	-1
Γ^3	2	-1	0

We see that now the whole table satisfies all the orthogonality relations.

(b) <u>Abelian Groups</u>. Since number of classes = number of elements = h, and number of representations = number of classes, it follows that there are h irreducible representations. Hence, $\overset{h}{\underset{i}{\Sigma}}\ell_i^2 = h$ implies $\ell_i = 1$ for all i. Hence every irreducible representation has dimension 1. This fact has several immediate consequences.

(i) Matrix multiplication becomes ordinary multiplication.

(ii) Unitarity of all representations means in this case that the numbers representing the abelian groups have modulus 1. Thus all the numbers lie on the unit circle.

(iii) If the number A represents the element a, then A^* represents a^{-1}. e is represented by 1.

(iv) The characters and the representations are the same thing. The character table is also a representation table.

(v) Ordinary multiplication of characters must be homomorphic with the group product of the elements.

Example: Non-cyclic groups of 4th order [isomorphic to the point groups D_2 or 222, C_{2v} or 2mm, and C_{2h} or 2/m]

	e	a	b	c
e	e	a	b	c
a	a	e	c	b
b	b	c	e	a
c	c	b	a	e

We find the character table as follows. The unit representation gives the first row. The dimensionality gives the first column. e must be represented everywhere by 1. $a^2 = b^2 = c^2 = e$, so a, b, and c can only be represented by ± 1. If a is represented by +1, then b and c must both be either +1 or -1 since bc = a. This gives Γ^2. If a is -1, then either b = +1 and c = -1 (Γ^3) or b = -1 and c = +1 (Γ^4).

	e	a	b	c
Γ^1	1	1	1	1
Γ^2	1	1	-1	-1
Γ^3	1	-1	1	-1
Γ^4	1	-1	-1	1

We might also have derived the table from the three invariant subgroups, $S_a = \{e,a\}$, $S_b = \{e,b\}$, $S_c = \{e,c\}$, and corresponding cosets in the three possible second order factor groups.

(c) <u>Cyclic Groups</u>. These are abelian groups which can be completely generated from the h powers of a single element. $G = \{e = a^h, a, a^2, a^3 \ldots a^{h-1}\}$. Now if we represent a by a number A_k, it follows that we must represent a^2 by A_k^2, etc. Thus e must be represented by A_k^h. But we know that e is represented by 1. So A_k must be one of the h h^{th} roots of unity.

$$A_k = e^{i \cdot 2\pi k/h}, \text{ and the character table is}$$

	e	a	a^2	...	a^{h-1}
Γ^1	1	1	1	...	1
Γ^2	1	A_1	A_1^2	...	A_1^{h-1}
Γ^3	1	A_2	A_2^2	...	A_2^{h-1}
$\vdots$	$\vdots$	$\vdots$	$\vdots$		$\vdots$
Γ_h	1	A_{h-1}	A_{h-1}^2	...	A_{h-1}^{h-1}

(i) <u>Order 2</u>. We have already used this table for the factor group of the group of 3-fold rotations of example A. [This group is isomorphic with point groups C_2 or 2, C_{1h} or m, and S_2 or $\bar{1}$.]

	e	a
Γ^1	1	1
Γ^2	1	-1

(ii) Order 3. [Isomorphic with C_3 or 3.]

$$\omega = \exp\left(\frac{2\pi i}{3}\right) = -\tfrac{1}{2} + \frac{i\sqrt{3}}{2}$$

	e	a	a^2
Γ^1	1	1	1
Γ^2	1	ω	ω^2
Γ^3	1	ω^2	ω

(iii) <u>Order 4</u>. [Isomorphic with C_4 or 4 and S_4 or $\bar{4}$.]

$\Gamma^4 = \Gamma^{3*}$

	e	a	a^2	a^3
Γ^1	1	1	1	1
Γ^2	1	-1	1	-1
Γ^3	1	i	-1	-i
Γ^4	1	-i	-1	i

(iv) <u>Order 6</u>. [Isomorphic with point groups C_6 or 6, C_{3h} or $\bar{6}$, and S_6 or $\bar{3}$.]

$\Gamma^4 = \Gamma^{3*}$
$\Gamma^6 = \Gamma^{5*}$

	e	a	a^2	a^3	a^4	a^5
Γ^1	1	1	1	1	1	1
Γ^2	1	-1	1	-1	1	-1
Γ^3	1	ω	ω^2	1	ω	ω^2
Γ^4	1	ω^2	ω	1	ω^2	ω
Γ^5	1	$-\omega$	ω^2	-1	ω	$-\omega^2$
Γ^6	1	$-\omega^2$	ω	-1	ω^2	$-\omega$.

(d) <u>Group of the Tetrahedron [T or 23]</u>. It is a group of order 12, and it was given as an example at the end of Chapter I (page 14). There are four classes and consequently four irreducible representations. The rela - tion

$$\ell_1{}^2 + \ell_2{}^2 + \ell_3{}^2 + \ell_4{}^2 = 12$$

is satisfied only by $\ell_1 = \ell_2 = \ell_3 = 1$, $\ell_4 = 3$. Thus

	$\mathcal{C}_e$	$4\mathcal{C}_a$	$4\mathcal{C}_b$	$3\mathcal{C}_c$
Γ^1	1	1	1	1
Γ^2	1			
Γ^3	1			
Γ^4	3			

There is an invariant subgroup $S_4 = \mathbf{C}_e + \mathbf{C}_c$ of index 3, and the factor group is the third order cyclic group whose irreducible representations have just been given. We obtain then three representations by associating with $S_4 = \{\mathbf{C}_e, \mathbf{C}_c\}$ the numbers 1,1,1; with $\mathbf{C}_a$: $1, \omega, \omega^2$; with $\mathbf{C}_b$: $1, \omega^2, \omega$. The first set reproduces Γ^1, and the other two give Γ^2 and Γ^3.

	$\mathbf{C}_1$	$4\mathbf{C}_a$	$4\mathbf{C}_b$	$3\mathbf{C}_c$
Γ^1	1	1	1	1
Γ^2	1	ω	ω^2	1
Γ^3	1	ω^2	ω	1
Γ^4	3	a	b	c

a, b, and c are determined by orthogonality with column 1.

$$1+\omega+\omega^2+3a = 0 \qquad a = 0$$
$$1+\omega^2+\omega+3b = 0 \qquad b = 0$$
$$1+1+1+3c = 0 \qquad c = -1$$

	$\mathbf{C}_1$	$4\mathbf{C}_a$	$4\mathbf{C}_b$	$3\mathbf{C}_c$
Γ^1	1	1	1	1
Γ^2	1	ω	ω^2	1
Γ^3	1	ω^2	ω	1
Γ^4	3	0	0	-1

THE REGULAR REPRESENTATION

Theorem VIII. If $\chi(R)$ is the character of a reducible representation of the element R, then it is possible to decompose it into irreducible characters according to $\chi(R) = \sum_i a_i \chi^{(i)}(R)$, where the a_i are non-negative integers whose values are determined from

$$a_i = \frac{1}{h} \sum_k \chi^{(i)}(C_k)^* \chi(C_k) N_k.$$

<u>Proof</u>: The first proof follows from the definition of the character as a trace and its invariance under the similarity transformations which bring the reducible representations into block form until they are irreducible. The sum of the diagonal elements is the sum of all such sums for the separate blocks and, further, a given irreducible representation can only occur either an integral number of times or be absent entirely.

To find a_i

$$\chi(R) = \sum_i a_i \chi^{(i)}(R)$$

and we must multiply by $\chi^j(R)^*$ and sum over h.

$$\sum_R \chi^{(j)}(R)^* \chi(R) = \sum_i a_i \sum_R \chi^{(j)}(R)^* \chi^{(i)}(R) = \sum_i a_i \delta_{ij}$$

by the first orthogonality relation. Collecting terms into classes immediately gives

$$\sum_k \chi^{(j)}(C_k)^* \chi(C_k) N_k = a_j h \qquad \text{Q.E.D.}$$

<u>The Regular Representation</u>. <u>Definition</u>: Form the group multiplication table in the following way. Establish the ordering of the columns by putting e first and the rest of the elements in some convenient order. Replace each element in the list by its inverse, and use the new list to establish the order of the rows. Then the table has the form

	e	α	β	γ	...
e	e				
α^{-1}		e			
β^{-1}			e		
γ^{-1}				e	

with e occurring along the diagonal.

We obtain the regular representation of an element α by forming an h x h matrix from the table by writing 1 wherever the element α occurs in the table and zero everywhere else.

Properties:

(i) It is a representation of the group. Each matrix is non-singular, since 1 appears once in each row and in each column, and so has an inverse. The unit element is the h x h unit matrix, which represents e. Suppose we label the group elements α_i, $i = 1,2,3,\ldots h$, and the corresponding matrices $A^{(i)}$ with elements $A_{jk}^{(i)}$. Then it is simple to replace the wordy definition of the regular representation given above by an algebraic one:

$$A_{jk}^{(i)} = 1 \quad \text{if } \alpha_j^{-1}\alpha_k = \alpha_i$$
$$= 0 \quad \text{otherwise.}$$

We next prove that

$$\sum_k A_{jk}^{(i)} A_{k\ell}^{(m)} = A_{j\ell}^{(p)}$$

if, and only if, $\alpha_i\alpha_m = \alpha_p$.

Proof: From the definition of $A_{jk}^{(i)}$,

$$A_{jk}^{(i)} \neq 0 \quad \text{only for k such that } \alpha_k = \alpha_j\alpha_i$$
$$A_{k\ell}^{(m)} \neq 0 \quad \text{only for k such that } \alpha_k = \alpha_\ell\alpha_m^{-1}.$$

It then follows that

$$\sum_k A_{jk}^{(i)} A_{k\ell}^{(m)} = 1 \quad \text{if, and only if, } \alpha_k = \alpha_j\alpha_i = \alpha_\ell\alpha_m^{-1}$$
$$= 0 \quad \text{otherwise;}$$

but

$$\alpha_j\alpha_i = \alpha_\ell\alpha_m^{-1}$$

implies

$$\alpha_j^{-1}\alpha_\ell = \alpha_i\alpha_m \equiv \alpha_p$$

which is the definition of $A_{j\ell}^{(p)}$.

(ii) The regular representation is a <u>faithful</u> representation. This means that the matrices are iso - morphic with the group or

$$A^{(i)} = A^{(j)} \quad \text{only if } \alpha_i = \alpha_j.$$

The proof is trivially obtained from the definition.

(iii) <u>Characters</u>. From the definition,

$$A_{jj}^{(i)} = 1 \quad \text{if } \alpha_i = \alpha_j^{-1}\alpha_j = e$$
$$= 0 \quad \text{otherwise.}$$

Consequently,

$$\chi^{reg}(\alpha_i) = \sum_{j=1}^{h} A_{jj}^{(i)} = h \quad \text{if } \alpha_i = e$$
$$= 0 \quad \text{otherwise.}$$

<u>Theorem IX</u>. The sum of the squares of the dimensions of all inequivalent representations is equal to the order of the group. $\sum \ell_i^2 = h$.

Proof: We apply Theorem VIII to the characters of the regular representation.

If

$$\chi^{(reg)}(R) = \sum_i a_i \chi^{(i)}(R),$$

then

$$a_i = \frac{1}{h} \sum_R \chi^{(i)}(R)^* \chi^{(reg)}(R)$$

But

$$\chi^{(i)}(e) = \ell_i, \quad \chi^{reg}(e) = h, \quad \chi^{reg}(a \neq e) = 0.$$

Thus

$$a_i = \ell_i.$$

Each irreducible representation appears in the regular representation a number of times equal to its dimension.

Since the dimension of the regular representation is h, and since it is equal to the sum of the dimensions of its irreducible components $(\sum_i a_i \ell_i)$, it follows that $\sum \ell_i^2 = h$.

Q.E.D.

III

Relationship to Quantum Mechanics

Before going ahead with the formal development of group theory, let us stop for a time and try to give motivation to the work we have already done and to the theorems to be proved in the future.

The fundamental problem in quantum mechanics is the following: Given the Hamiltonian operator of a system, find its complete set of eigenfunctions and eigenvalues. A typical solution of the problem begins with the construction of a Hamiltonian matrix by means of some arbitrarily chosen set of functions that spans the space of the operator. Then the worker proceeds to diagonalize the matrix.

Now in any but the simplest systems an exact solution is out of the question because of the difficulty of the diagonalization. Instead the worker displays all of the cleverness at his disposal in his choice of the basis set of functions used to construct the Hamiltonian matrix; he diagonalizes as much of it as he can; and then he argues as convincingly as he can that the contributions of the remaining non-diagonal terms are so small that they can be neglected.

But before the last step, which seems invariably to be accompanied by some waving of hands, he is happy to discover as many <u>rigorously exact</u> simplifications as the problem will allow. Let us suppose, for example, that the Hamiltonian operator of a system, which is a functional of the coordinates of the system, remains invariant under spatial inversion: $(x,y,z) \rightarrow (-x,-y,-z)$. Then it is possible to define an operator $\hat{P}$ such that $\hat{P}\ \psi(\underline{x}) = \psi(-\underline{x})$. It is easy to prove that $\hat{P}$ is linear and hermitian and that it evidently commutes with the Hamiltonian. If $\Phi(\underline{x})$ is an eigenfunction of $\hat{P}$, that is, if $\hat{P}\Phi = p\Phi$, then $\hat{P}(\hat{P}\Phi) = \hat{P}(p\Phi) = p^2\Phi$. But $\hat{P}^2$ is only the identity operator, so $p = \pm 1$.

Thus $\hat{P}$ has two distinct eigenvalues, ± 1, the even and odd parities. Its eigenfunctions are the even and the odd functions in the function space under consideration.

Now let us construct the matrices of H and $\hat{P}$ using functions which span the space of $\hat{H}$ and are also eigenfunctions of $\hat{P}$. Next we work, in matrix notation, the commutativity of $\hat{H}$ and $\hat{P}$ noting that P is now diagonal.

$$(PH)_{ik} = (HP)_{ik}$$

$$\sum_j P_{ij} H_{jk} = \sum_\ell H_{i\ell} P_{\ell k} \quad \text{But } P_{rs} = P_{rr} \delta_{rs}$$

$$\text{so} \quad P_{ii} H_{ik} = H_{ik} P_{kk} \quad \text{or} \quad (P_{ii} - P_{kk}) H_{ik} = 0$$

Hence we must conclude that $H_{ik} = 0$ whenever $P_{ii} \neq P_{kk}$. We say, in short, that there are no matrix elements of $\hat{H}$ between states of different parity.

And so by this choice of basis functions of definite parity, we have accomplished a partial diagonalization of the Hamiltonian matrix without any detailed knowledge of the operator. It seems clear that a knowledge of all operations which commute with the Hamiltonian of a system and a knowledge of their eigenfunctions and eigenvalues would be of considerable help in simplifying any problem in quantum mechanics. And it is the theory of groups which provides a basis for the study of these so-called "symmetry operations of the Hamiltonian". Its objective is a suitable generalization of some of the following characteristics of the parity operation:

(1) $\hat{P}$ is a linear, unitary operation.

(2) $\hat{P}\hat{H} = \hat{H}\hat{P}$, if ψ is an eigenfunction of $\hat{H}$, then $\hat{P}\psi$ is degenerate with it.

(3) The identity operation $\hat{E}$ and $\hat{P}$ form a group. It is 2nd order, cyclic. The representations are one-dimensional, and the character table, which is also a representation table, is given by

	$\hat{E}$	$\hat{P}$
Γ^e	1	1
Γ^o	1	-1

(4) If Φ_e and Φ_o are even and odd eigenfunctions of $\hat{P}$, then $\hat{E}\Phi_e = \Phi_e$, $\hat{P}\Phi_e = \Phi_e$, $\hat{E}\Phi_o = \Phi_o$, and $\hat{P}\Phi_o = -\Phi_o$. In short, $\hat{R}\Phi_e = \Gamma^e(\hat{R})\Phi_e$ and $\hat{R}\Phi_o = \Gamma^o(\hat{R})\Phi_o$, where $\hat{R}$ stands for $\hat{E}$ or $\hat{P}$. We may say that "Φ_e transforms under the group $\{\hat{E},\hat{P}\}$ according to the representation Γ^e " and similarly for Φ_o.

(5) Φ_e and Φ_o are orthogonal.

(6) $H_{eo} \equiv \langle \Phi_e \mid \hat{H}\Phi_o \rangle = 0$

(7) $\int \hat{P}f(\underline{x})d^{3N}x = \int f(\underline{x})d^{3N}x$ if boundary surface is infinite or if it has inversion symmetry.

(8) Every function in the space of $\hat{H}$ can be expressed as a linear combination of even and odd functions: $\psi(x) = \frac{\psi + \hat{P}\psi}{2} + \frac{\psi - \hat{P}\psi}{2}$. Thus it is possible to choose eigenfunctions of a degenerate state as purely even or purely odd.

With this as an introduction, the outlines of our task are clear. Now we must develop general theorems concerning the symmetry operations of the Hamiltonian — theorems which contain as special cases the results we know for the parity operation.

SYMMETRY TRANSFORMATIONS

Definition: A symmetry transformation with respect to a given Hamiltonian is a linear coordinate transformation such that the form of the Hamiltonian operator is the same in the old and the new coordinate system.

Examples: The following classes of linear transformations are frequently symmetry transformations. They are all real.

(1) Translations: $\underline{x}' = \underline{x} + \underline{a}$ (or $x_i' = x_i + a_i$)

(2) Rotations: $\underline{x}' = R\underline{x}$ (or $x_i' = \sum_j R_{ij} x_j$), $|R| = 1$.

(3) Inversion: $\underline{x}' = -\underline{x}$ (or $x_i' = -x_i$)

(4) Reflections: $\underline{x}' = R\underline{x}$, $|R| = -1$. We can arrive at every reflection by a proper rotation and an inversion.

(5) Permutations: $\underline{x}_1' = \underline{x}_2$, $\underline{x}_2' = \underline{x}_1$

All these transformations contain an inverse transformation. The proper and improper rotations are real, unitary transformations. Thus they are orthogonal transformations $R^{-1} = R^\dagger = \widetilde{R}^* = \widetilde{R}$, or $(R^{-1})_{ij} = R_{ji}$. Any set of these transformations which satisfies closure is a group, if it contains the unit (or identity) and inverse elements

Definition: The transformation operator $\hat{P}_R$ associated with the symmetry transformation R is defined by the following operator equation which must be an identity in $\underline{x}$.

$$\hat{P}_R f(\underline{x}) \equiv f(R^{-1}\underline{x})$$

It must be emphasized that the operator acts upon the coordinates $\underline{x}$ and not on the argument of f. Thus we mean that $\hat{P}_R f(S\underline{x}) \equiv f(SR^{-1}x)$ and $\neq f(R^{-1}S\underline{x})$.

Theorem: If $\hat{P}_R\hat{P}_S$ is defined to mean "$\hat{P}_S$ operates first and then $\hat{P}_R$ operates," then the set $\{\hat{P}_R\}$ is isomorphic with the set $\{R\}$.

Proof:

$$\hat{P}_R\hat{P}_S f(\underline{x}) = \hat{P}_R(f[S^{-1}\underline{x}]) = f(S^{-1}R^{-1}\underline{x})$$

$$= f([RS]^{-1}\underline{x}) = \hat{P}_{(RS)} f(\underline{x})$$

Hence $\hat{P}_R\hat{P}_S = \hat{P}_T$ if, and only if, RS = T   Q.E.D

Theorem: If $\{R\}$ is the set of all symmetry transformations of a Hamiltonian, then $\hat{P}_R$ commutes with $\hat{H}$ for all R and the set $\{\hat{P}_R\}$ is a group, called the group of the

Schrödinger equation.

Proof: First we observe that if R is a symmetry transformation, then so is R^{-1}; for if $\hat{H}$ is invariant when written in terms of $\underline{x}' = R\underline{x}$, then it will surely be invariant when written in terms of $\underline{x}'' = R^{-1}\underline{x}'$ because $\underline{x}'' = \underline{x}$. The identity transformation $\underline{x}' = I\underline{x} = \underline{x}$ is obviously a symmetry transformation, and a succession of two symmetry transformations is itself a symmetry transformation. Hence the set of all symmetry transformations of a given Hamiltonian, and also the transformation operators iso-morphic to it, makes up a group.

The commutativity of $\hat{P}_R$ and $\hat{H}$ will appear obvious from the following considerations. R is defined through the relation $\underline{x}' = R\underline{x}$. The operation $\hat{P}_R$ may be thought of as a two-step process. First, write the operand in terms of the new coordinates by solving the equation for $\underline{x}$. That is, replace $\underline{x}$ by its equivalent $R^{-1}\underline{x}'$. Second, relabel the new variable $\underline{x}$. Now it is obvious that $\hat{P}_R$ has no effect on $\hat{H}$ if R is a symmetry transformation, by very definition. Since $\hat{P}_R$ acts only on the operand of $\hat{H}$, it commutes with $\hat{H}$.

DEGENERACY AND INVARIANT SUBSPACES

Normal Degeneracy. Suppose E_n is a g-fold degenerate energy level with linearly independent eigenfunctions $\{\psi_\varkappa\}$, $\varkappa = 1, 2 \ldots g$. Then $\hat{H}\psi_\varkappa = E_n\psi_\varkappa$. If $\hat{P}_R$ is a sym-metry operator of $\hat{H}$, then $\hat{P}_R(\hat{H}\psi_\varkappa) = \hat{P}(E_n\psi_\varkappa)$ and $\hat{H}(\hat{P}_R\psi_\varkappa) = E_n(\hat{P}_R\psi_\varkappa)$. Thus $\hat{P}_R\psi_\varkappa$ must be an eigenfunction of H. If we think of $\{\psi_\varkappa\}$ as spanning a certain g-dimensional vector space, then $\hat{P}_R\psi_\varkappa$ must be a vector in that space for all R in G, the group of the Schrödinger equation, and for $\varkappa = 1$, $\ldots$ g. This is true because $\hat{P}_R\psi_\varkappa$ must be a linear combin-ation of the $\{\psi_\varkappa\}$.

Suppose that we fix a value for $\varkappa$ and form the set of eigenfunctions $\{\hat{P}_R\psi_\varkappa\}$ for all R in G. If this set spans the same vector space as that of the set $\{\psi_\varkappa\}$, $\varkappa = 1 \ldots g$, then the degeneracy is termed normal. If on the other hand

it spans only a subspace of the complete space associated
with the eigenfunctions of E_n, then the separate subspaces
are termed <u>accidentally</u> degenerate.

These are definitions. How in practice do we decide whether we have normal or accidental degeneracy? First
we find all the symmetry operations of the Hamiltonian
which form a group, and then we test whether the set of
functions $\hat{P}_R \psi_\varkappa$, for given $\varkappa$, spans the entire space. If
it does, then the degeneracy is normal. If not, it may
still be normal, for the group of operators actually used
may be merely a subgroup of G. We may not have found all
of the symmetry operations of $\hat{H}$. Hence we are never certain that a degeneracy is indeed accidental. In ordinary
usage, therefore, "accidental degeneracy" always refers
to a stated group of symmetry operations, which may or may
not be the entire group of the Schrödinger equation. In
this latter sense we may say that normal degeneracy is a
consequence of the stated symmetries of the Hamiltonian,
although accidental degeneracy is not. Thus it may be
possible to remove an accidental degeneracy by changing
a parameter of the Hamiltonian, a coupling constant, say,
which does not alter the stated symmetry.

It is useful at this point to define an <u>invariant
subspace</u>.

<u>Definition:</u> A set $\{\psi_i\}$ of vectors is said to span
an invariant subspace V_S under a given set of operations
$\{\hat{P}_j\}$ if $\hat{P}_j \psi_i$ is also in V_S for all j and i. (Note that
$P_j \psi_i$ need not be one of the basis vectors of V_S in order
to be contained in V_S; it need only be a linear combina-
tion of them.) For example, the column vectors $\begin{pmatrix} 1 \\ 0 \\ 0 \\ 0 \end{pmatrix}$ and $\begin{pmatrix} 0 \\ 1 \\ 0 \\ 0 \end{pmatrix}$
under operations of matrix multiplication by

$$\begin{pmatrix} a & b & 0 & 0 \\ c & d & 0 & 0 \\ 0 & 0 & 1 & 0 \\ 0 & 0 & 0 & 1 \end{pmatrix}$$

for any a, b, c, d span an invariant subspace.

In this language, therefore, we have shown that all degenerate eigenfunctions of a particular energy span an invariant subspace under all symmetry operations of the Hamiltonian.

Definition: If a subspace is invariant under $\{\hat{P}_J\}$ and if it can be divided into smaller subspaces also invariant under $\{\hat{P}_J\}$, then the larger subspace is termed "reducible"; otherwise it is called "irreducible." For example, if $b = c = 0$ in the matrix above, then the first or second column vector would span an invariant subspace within that spanned by both of the vectors and the larger space would be reducible. On the other hand, if b and c cannot be put equal to zero in all of the matrices, then the larger space is irreducible. The reducibility of this vector space evidently depends upon the reducibility of the matrix operators in question.

In this language, therefore, we may say that we have normal degeneracy if all of the eigenfunctions of a particular energy span an irreducible invariant subspace. If the subspace is reducible, the degeneracy is termed accidental.

Theorems concerning invariant subspaces. These will be immediately applicable to subspaces of eigenfunctions of a Hamiltonian, but the proofs are stated for any subspace V_S spanned by the set of vectors $\{\phi_\mu\}$ which is invariant under a **group** G of operations $\{\hat{P}_R\}$. We shall often have occasion to use arbitrary functions which span invariant subspaces, and these theorems apply there as well.

When we say that V_S is an invariant subspace, we mean that for all $\hat{P}_R$ in G we can find a matrix $\Gamma(R)$ of scalar coefficients such that $\hat{P}_R\phi_\mu = \sum_\lambda \phi_\lambda \Gamma(R)_{\lambda\mu}$, where the sum is extended over all the g linearly independent vectors which span V_S. If we think of the set $\{\phi_\mu\}$ as forming a row vector $\underline{\phi}$, then we may also use the more compact vector notation:

$$\hat{P}_R\underline{\phi} = \underline{\phi}\ \Gamma(R)$$

(i) The set $\{\Gamma(R)\}$ is a representation of G.

Proof:

$$\hat{P}_S(\hat{P}_R\underline{\phi}) = \hat{P}_S\underline{\phi}\ \Gamma(R)$$

$$(\hat{P}_S\hat{P}_R)\underline{\phi} = \underline{\phi}\Gamma(S)\Gamma(R)$$

Now if $\hat{P}_S\hat{P}_R = \hat{P}_T$ then $\hat{P}_T\underline{\phi} = \underline{\phi}\Gamma(S)\Gamma(R)$. But $P_T\underline{\phi} = \underline{\phi}\Gamma(T)$. Hence $\Gamma(S)\Gamma(R) = \Gamma(T)$. Thus the set of matrices $\{\Gamma(R)\}$ under matrix multiplication is homomorphic with G and so is a representation of it.

(ii) A change of basis vectors $\psi_{\varkappa} = \Sigma_{\lambda}\phi_{\lambda}a_{\lambda\varkappa}$ or more compactly, $\underline{\psi} = \underline{\phi}A$ amounts to nothing more than an equivalence transformation on Γ.

Proof: $\hat{P}_R\underline{\phi} = \underline{\phi}\Gamma(R)$ and $\underline{\phi} = \underline{\psi}A^{-1}$. Thus

$$\hat{P}_R(\underline{\psi}A^{-1}) = (\underline{\psi}A^{-1})\Gamma(R)$$

$$(\hat{P}_R\underline{\psi})A^{-1} = \underline{\psi}\ [A^{-1}\Gamma(R)]$$

$$\hat{P}_R\underline{\psi} = \underline{\psi}[A^{-1}\Gamma(R)A] \equiv \underline{\psi}\Gamma'(R)$$

Since A is independent of R, the transformation is the same for all R.

(iii) If an inner product between basis vectors is defined [and is symbolized by $(\phi_{\varkappa}\phi_{\lambda}) \equiv J_{\varkappa\lambda}$ or, in our compact notation $(\underline{\tilde{\phi}},\underline{\phi}) = J$], then it is always possible to choose mutually orthogonal normalized basis vectors, that is $J_{\varkappa\lambda} = \delta_{\varkappa\lambda}$ or $J = I$. This we do not prove but go on to show that if the basis functions are orthonormal, then $\Gamma(R)$ is the matrix, in the quantum mechanical sense, of the operator $\hat{P}_R$.

Proof:

$$\hat{P}_R\underline{\phi} = \underline{\phi}\Gamma(R)$$

$$(\underline{\tilde{\phi}},\hat{P}_R\underline{\phi}) = (\underline{\tilde{\phi}},\underline{\phi}\Gamma(R)) = (\underline{\tilde{\phi}},\underline{\phi})\Gamma(R) = \Gamma(R)$$

(iv) If the basis functions are orthonormal and $\hat{P}_R$ is a unitary operator, then $\Gamma(R)$ is a unitary matrix, if the following identity is satisfied for all scalars a and b: $\left(a\phi_\varkappa, b\phi_\lambda\right) \equiv a^*b\left(\phi_\varkappa, \phi_\lambda\right)$.

Proof: A unitary operator is one that preserves inner products.

$$\left(\widetilde{P_R\phi}, P_R\phi\right) = \left(\underline{\widetilde{\phi}}, \underline{\phi}\right)$$

But $\left(\underline{\widetilde{\phi}}, \underline{\phi}\right) = I$, so

$$I = \left(\widetilde{\underline{\phi}\Gamma}, \underline{\phi}\Gamma\right) = \left(\widetilde{\Gamma}\widetilde{\phi}, \phi\Gamma\right) = \widetilde{\Gamma}^*\left(\underline{\widetilde{\phi}}, \underline{\phi}\right)\Gamma = \Gamma^\dagger I\Gamma = \Gamma^\dagger\Gamma.$$

(v) If an invariant subspace V_S is irreducible under all the operations of G, then the set $\{\Gamma(R)\}$ is an irreducible representation of G. It is sufficient to prove the contra-positive. Suppose $\{\Gamma(R)\}$ is reducible by the equivalence transformation A, which is the same for all R. That is, $\Gamma'(R) = A^{-1}\Gamma(R)A$ is in block form. Then, by theorem (ii), the new basis vectors $\underline{\psi} = \underline{\phi}A$ must transform according to the equation $\hat{P}_R\underline{\psi} = \underline{\psi}\Gamma'(R)$. But since $\Gamma'(R)$ is in block form for all R, we have effected a reduction of V_S. Hence V_S is reducible. Q.E.D. If the particular irreducible representation is $\Gamma^{(i)}(R)$, we label the vectors of V_S, $\underline{\phi}^{(i)}$ and we say that they transform according to the ith irreducible representation of G.

(vi) If ϕ is an arbitrary vector in some vector space, then the set of h (not necessarily linearly independent) vectors $\{\hat{P}_R\phi\}$ formed from all $\hat{P}_R$ in G defines a subspace V_S invariant under G.

Proof: We must prove that $\hat{P}_S(\hat{P}_R\phi)$ is in V_S for all R and S. But $\hat{P}_S(\hat{P}_R\phi) = (\hat{P}_S\hat{P}_R)\phi$, and $\hat{P}_S\hat{P}_R$ is some $\hat{P}_T$ in G. Now V_S is defined to contain $\hat{P}_T\phi$. Hence $\hat{P}_S(\hat{P}_R\phi)$ is in V_S. Q.E.D. We do not show it here, but it can be proved that, if the h vectors are linearly independent, the representation generated from them by G is the regular representation.

(vii) If $\underline{\phi}^{(i)}$ is the set of ℓ_i orthogonal vectors spanning the subspace V_{Si} and transforming according to the irreducible representation $\Gamma^{(i)}(R)$ of a group G of unitary operations, and if $\underline{\Phi}^{(j)}$ is defined similarly, then all vectors in V_{Si} are orthogonal to all those in V_{Sj} when $i \neq j$.

Proof: It is sufficient to prove that the basis vectors in V_{Si} are orthogonal to those in V_{Sj}. We must show that $(\underset{\sim}{\widetilde{\Phi}}{}^{(i)}, \underline{\Phi}^{(j)})$ is an $\ell_i \times \ell_j$ null matrix if $i \neq j$.

$$(\underline{\widetilde{\Phi}}^{(i)}, \underline{\Phi}^{(j)}) = (\widehat{P}_R \underline{\Phi}^{(i)}, \widehat{P}_R \underline{\Phi}^{(j)}) \quad \text{by unitarity}$$

$$= (\underline{\Phi}^{(i)} \Gamma^{(i)}(R), \underline{\Phi}^{(j)} \Gamma^{(j)}(R)) \quad \begin{matrix} \text{by} \\ \text{hypothesis} \end{matrix}$$

$$= \Gamma^{(i)}(R)^\dagger (\underline{\widetilde{\Phi}}^{(i)}, \underline{\Phi}^{(j)}) \Gamma^{(j)}(R) \begin{matrix} \text{definition} \\ \text{of inner} \\ \text{product} \end{matrix}$$

$$= \Gamma^{(i)}(R)^{-1} (\underline{\widetilde{\Phi}}^{(i)}, \underline{\Phi}^{(j)}) \Gamma^{(j)}(R) \begin{matrix} \text{by Theo-} \\ \text{rem (iv)} \end{matrix}$$

Hence

$$\Gamma^{(i)}(R) (\underline{\widetilde{\Phi}}^{(i)}, \underline{\Phi}^{(j)}) = (\underline{\widetilde{\Phi}}^{(i)}, \underline{\Phi}^{(j)}) \Gamma^{(j)}(R)$$

Thus $(\underline{\widetilde{\Phi}}^{(i)}, \underline{\Phi}^{(j)})$ is a matrix which satisfies the postulates of Theorem III, page 26, and so it must be the null matrix when $i \neq j$. Q.E.D.

(viii) Any vector Φ in the space in which all of the operations of G are internal operations may be decomposed into a linear combination of component vectors each of which, in turn, has components only in a particular invariant irreducible subspace. In other words, given Φ, we can always find a set $\{\underline{\Phi}^{(i)}\}$ such that

$$\Phi = \sum_i \sum_{\varkappa=1}^{\ell_i} \Phi_\varkappa^{(i)} c_\varkappa^{(i)} \quad \text{and} \quad \widehat{P}_R \Phi_\varkappa^{(i)} = \sum_\lambda \Phi_\lambda^{(i)} \Gamma^{(i)}(R)_{\lambda \varkappa}$$

Proof: We form the set of vectors $\{\widehat{P}_R \Phi\}$ for all R in G. Then by Theorem (vi) these vectors span an invariant subspace. From them we take the largest set of linearly independent vectors and form an ortho-normal basis set and the representation generated by it. If the representation is irreducible, then we have found the $\{\Phi_{\varkappa}^{(i)}\}$. If not, then some similarity transformation will reduce it and the same transformation matrix will give us the basis vectors of the irreducible spaces it contains. Once all basis vectors are known, the coefficients $c_{\varkappa}^{(i)}$ can be found as follows:

$$(\Phi_{\lambda}^{(j)}, \Phi) = \sum_{i\varkappa} (\Phi_{\lambda}^{(j)}, \Phi_{\varkappa}^{(i)}) c_{\varkappa}^{(i)} = \sum_{i\varkappa} \delta_{ij} \delta_{\varkappa\lambda} c_{\varkappa}^{(i)} = c_{\lambda}^{(j)}$$

Application of these theorems to quantum mechanics. The invariant subspace is the set of degenerate eigenfunctions of a given Hamiltonian and a given energy. The unitary group of operations is the group of the Schrödinger equation. The eigenfunctions are normally degenerate if the subspace is irreducible. The inner product is defined by $(\Phi_{\varkappa}, \Phi_{\lambda}) \equiv \int \Phi_{\varkappa}^{*}(\underline{x}) \Phi_{\lambda}(\underline{x}) d^{3N}x$. Then theorems (i) through (viii) may be interpreted as follows:

(i) The eigenfunctions of a given energy generate a representation of G.

(ii) A linear transformation to new eigenfunctions generates a representation equivalent to the original.

(iii) If the eigenfunctions are ortho-normal, then $\Gamma(R)$ is merely the matrix of the operator $\hat{P}_R$.

(iv) The representations so generated are unitary.

(v) If the degeneracy is normal, the representation is irreducible.

(vi) An arbitrary function in the space of $\hat{H}, \Phi$ can be used to construct an invariant subspace by forming the operation $\hat{P}_R \Phi$ for all R in G.

(vii) Functions which transform according to two different irreducible representations of G are orthogonal. If the functions are, in particular, accidentally degenerate eigenfunctions, then the matrix element of $\hat{H}$ between them is zero.

(viii) Any function in the space of $\hat{H}$ can be decomposed into a linear combination of functions transforming according to irreducible representations of G.

A glance back at page 48 will confirm that each of these results is indeed a generalization of some of the characteristics of eigenfunctions which transform according to the irreducible representations of the parity group.

PROJECTION OPERATORS

Suppose the set $\{\psi_{\varkappa}^{(j)}\}$ are basis vectors of an invariant subspace transforming according to the jth irreducible representation of G. That is

$$\hat{P}_R \psi_{\varkappa}^{(j)} = \sum_\lambda \psi_\lambda^{(j)} \Gamma^{(j)}(R)_{\lambda\varkappa}$$

Multiply by $\Gamma^{(i)}(R)^*_{\lambda'\varkappa'}$ and sum the result over all R in G.

$$\sum_R \Gamma^{(i)}(R)^*_{\lambda'\varkappa'} \hat{P}_R \psi_{\varkappa}^{(j)} = \sum_{R\lambda} \psi_\lambda^{(j)} \Gamma^{(j)}(R)_{\lambda\varkappa} \Gamma^{(i)}(R)^*_{\lambda'\varkappa'}$$

We may perform the sum over R on the right side, making use of the orthogonality theorem, page 29

$$\sum_R \Gamma^{(i)}(R)^*_{\lambda'\varkappa'} \hat{P}_R \psi_{\varkappa}^{(j)} = \sum_\lambda \psi_\lambda^{(j)} \frac{h}{\ell_i} \delta_{ij} \delta_{\lambda\lambda'} \delta_{\varkappa\varkappa'} = \psi_{\lambda'}^{(j)} \frac{h}{\ell_i} \delta_{ij} \delta_{\varkappa\varkappa'}$$

Definition: Let $\hat{\rho}_{\lambda'\varkappa'}^{(i)} \equiv \frac{\ell_i}{h} \sum_R \Gamma^{(i)}(R)^*_{\lambda'\varkappa'} \hat{P}_R$. Then,

(i) $\qquad \hat{\rho}_{\lambda'\varkappa'}^{(i)} \psi_{\varkappa}^{(j)} = \delta_{ij} \delta_{\varkappa\varkappa'} \psi_{\lambda'}^{(j)}$

and $\qquad \rho_{\varkappa\varkappa}^{(i)} \psi_{\varkappa}^{(i)} = \psi_{\varkappa}^{(i)}$

Thus $\psi_{\varkappa}^{(i)}$ is an eigenfunction of $\hat{\rho}_{\varkappa\varkappa}^{(i)}$ with eigenvalue 1.

Definition: Let $\hat{\rho}^{(i)} \equiv \sum_\varkappa \hat{\rho}_{\varkappa\varkappa}^i$. Then we see that

$$\boxed{\hat{\rho}^{(i)} \equiv \frac{\ell_i}{h} \sum_R \chi^{(i)}(R)^* \hat{P}_R}$$

If we set $\lambda' = \varkappa'$ in equation (i) above and sum over $\varkappa'$, we get

$$\sum_{\kappa', \kappa', \kappa} \rho^{(i)}_{\kappa', \kappa} \psi^{(j)}_{\kappa} = \delta_{ij} \sum_{\kappa'} \delta_{\kappa\kappa'} \psi^{(j)}_{\kappa'}$$

$$\boxed{\hat{\rho}^{(i)} \psi^{(j)}_{\kappa} = \delta_{ij} \psi^{(j)}_{\kappa}}$$

Note that $\hat{\rho}^{(i)}$ is identical for all equivalent representations. Its eigenvalues are zero and one, and its effect is to project the components which transform according to the ith irreducible representation of G out of an arbitrary vector.

EXAMPLES OF TRANSFORMATIONS OF FUNCTIONS

We now consider some functions which may or may not be eigenfunctions of a Hamiltonian. We build from each of them, using Theorem (vi) page 54, an invariant subspace. We therefore generate, according to Theorem (i) page 53, a representation of the group in question. If it is reducible, we seek the irreducible components of the particular function, which components, according to Theorem (viii) page 55, must be present. In some instances we shall use the projection operators to project them out.

Since the groups employed in these examples contain only proper and improper rotations, $f(r) \equiv f(\sqrt{x^2+y^2+z^2})$ will always transform into itself. Hence it can only generate the identity representation of any of these groups. The functions employed in the examples shall be $f(r), xf(r), yf(r), zf(r), x^2 f(r)$, and $f(r)$ multiplied by the other quadratic forms. But since $xf(r)$ transforms exactly like x, and similarly for the others, we need only look at the representations generated by 1, x, **y**, z, x^2, etc. (1 is trivial and is omitted.)

Example A: The first group we shall consider is the so-called D_2 or 222 group which we have discussed on page 39. Its elements {eabc} are in this case E, the identity and three two-fold rotations about the cartesian axes: C_{2z}, C_{2y}, C_{2x}, respectively. The group is abelian; so the representations must be one dimensional. This means that every function must be transformed into some constant times itself. Every element is also its own inverse. It is always wise to start with the functions x, y, z taken collectively, since linear coordinate transformations of any type will only lead to linear functions. Thus $\hat{P}_E(x,y,z)=(x,y,z)$, $\hat{P}_{C_{2z}}(x,y,z)=(-x,-y,z)$, $\hat{P}_{C_{2y}}(x,y,z) = (-x,y,-z)$, $\hat{P}_{C_{2x}}(x,y,z) = (x,-y,-z)$. Now we write the character table for the D_2 group, which is also a representation table since the group is abelian; and we write out a table of transformations for the functions.

	E	C_{2z}	C_{2y}	C_{2x}
A	1	1	1	1
B_1	1	1	-1	-1
B_2	1	-1	1	-1
B_3	1	-1	-1	1

	E	C_{2z}	C_{2y}	C_{2x}
x	x	-x	-x	x
y	y	-y	y	-y
z	z	z	-z	-z
x^2	x^2	x^2	x^2	x^2
xy	xy	xy	-xy	-xy
etc.				

And so we conclude that x^2, y^2, and z^2 transform according to the irreducible representation A, that z and xy transform according to B_1, that y and xz transform according to

B_2, and that x and zy transform according to B_3. If $xf(r)$, $yf(r)$, and $zf(r)$ were the degenerate p-functions of a free atom, and if the atom were placed in an external field with D_2 symmetry, then they would either split apart or be accidentally degenerate.

Example B: The second example is the group C_{2v} or 2mm, isomorphic with D_2 but physically different since it replaces two two-fold axes with two perpendicular mirror planes containing the z-axis. It is easy to write the transformations of (x,y,z) and so derive the transformation table:

	E	C_{2z}	σ_x	σ_y
A_1	1	1	1	1
A_2	1	1	-1	-1
B_1	1	-1	1	-1
B_2	1	-1	-1	1

	E	C_{2z}	σ_x	σ_y	
x	x	-x	-x	x	
y	y	-y	y	-y	
z	z	z	z	z	
x^2	x^2	x^2	x^2	x^2	
xy	xy	xy	-xy	-xy	
yz	yz	-yz	yz	-yz	etc.

Thus z, x^2, y^2 and z^2 transform according to A_1; xy, according to A_2; y and yz, according to B_1; x and xz according to B_2.

Example C: The third example is the group of the square, D_4 in the Schoenflies notation or 422 in the international notation. It is a group of order 8 consisting of e   the identity

a_x a rotation by π about the x-axis

a_y a rotation by π about the y-axis

d_1 a rotation by π about the axis x=y

d_2 a rotation by π about the axis x=-y

b   a rotation by π about the z-axis

c_1 a rotation by $\pi/2$ about the z-axis

c_2 a rotation by $3\pi/2$ about the z-axis

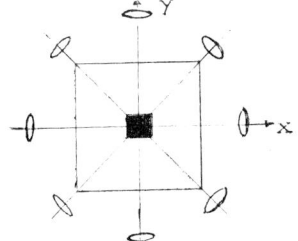

The group multiplication table is

$\dfrac{1}{2}$	e	a_x	a_y	d_1	d_2	b	c_1	c_2
e	e	a_x	a_y	d_1	d_2	b	c_1	c_2
a_x	a_x	e	b	c_2	c_1	a_y	d_2	d_1
a_y	a_y	b	e	c_1	c_2	a_x	d_1	d_2
d_1	d_1	c_1	c_2	e	b	d_2	a_x	a_y
d_2	d_2	c_2	c_1	b	e	d_1	a_y	a_x
b	b	a_y	a_x	d_2	d_1	e	c_2	c_1
c_1	c_1	d_1	d_2	a_y	a_x	c_2	b	e
c_2	c_2	d_2	d_1	a_x	a_y	c_1	e	b

The group is isomorphic with C_{4v} or $4mm$, and D_{2d} or $\bar{4}2m$

Classes:
$$\begin{aligned}
\mathcal{C}_e &= e \\
\mathcal{C}_a &= a_x \; a_y \\
\mathcal{C}_b &= b \\
\mathcal{C}_c &= c_1 \; c_2 \\
\mathcal{C}_d &= d_1 \; d_2
\end{aligned}$$

Invariant subgroups and cosets

$$S_1 = \{C_e, C_b, C_c\} \qquad C_1 = \{C_a, C_d\}$$

$$S_2 = \{C_e, C_b, C_a\} \qquad C_2 = \{C_c, C_d\}$$

$$S_3 = \{C_e, C_b, C_d\} \qquad C_3 = \{C_a, C_c\}$$

$$S_4 = \{C_e, C_b\}; C_4^1 = C_a; C_4^2 = C_c; C_4^3 = C_d.$$

Character Table

	C_e	$2C_a$	$2C_d$	C_b	$2C_c$
Γ_1	1	1	1	1	1
Γ_2	1	-1	-1	1	1
Γ_3	1	1	-1	1	-1
Γ_4	1	-1	1	1	-1
Γ_5	2	0	0	-2	0

In the previous examples each element was its own inverse. We must now be more careful since this is no longer true for the 4-fold rotations c_1 and c_2. Recall that $\hat{P}_R f(\underline{x}) \equiv f(R^{-1}\underline{x})$. So if R is a rotation by an angle θ, then R^{-1} is a rotation by $-\theta$. The element c_1 corresponds to an anticlockwise rotation of the square by $\pi/2$. Hence the inverse operation, the clockwise rotation by $\pi/2$, is to be applied to $\underline{x}$. That is

$$c_1(x,y,z) = (x' = -y, \; y' = x, \; z' = z)$$

$$c_1^{-1}(x,y,z) = c_2(x,y,z) = (x' = y, \; y' = -x, \; z' = z)$$

$$\hat{P}_{c1}f(\underline{x}) = \hat{P}_{c1}f(x,y,z) = f(c_2\underline{x}) = f(y,-x,z)$$

Thus as we see in the figure

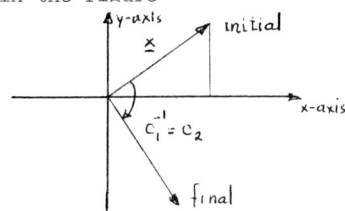

the x-component of the final state is the y-component of
the initial state; and the y-component of the final state
is minus the x-component of the initial state. There is
an alternative way of expressing this, but we should accept
one view or the other; holding both views simultaneously
may be in some instances most confusing. The second way
of expressing $\hat{P}_R$ is to say that the operation R is to be
carried out on the coordinate axes of $\underline{x}$. In this way P_{c1}
means

$$\text{final x-axis} = \text{initial y-axis}$$
$$\text{final y-axis} = \text{initial negative x-axis}$$
$$\text{final z-axis} = \text{initial z-axis}$$

This second view is perhaps simpler since we never need
consider the inverse operation. With these considerations,
the transformation table is

	e	a_x	a_y	b	c_1	c_2	d_1	d_2
x	x	x	$-x$	$-x$	y	$-y$	y	$-y$
y	y	$-y$	y	$-y$	$-x$	x	x	$-x$
z	z	$-z$	$-z$	z	z	z	$-z$	$-z$
x^2-y^2	x^2-y^2	x^2-y^2	x^2-y^2	x^2-y^2	y^2-x^2	y^2-x^2	y^2-x^2	y^2-x^2
xy	xy	$-xy$	$-xy$	xy	$-xy$	$-xy$	xy	xy

etc.

Comparing these results with the table of characters, we can confirm that z^2, $x^2 + y^2$ transform according to Γ_1; z transforms according to Γ_2; $x^2 - y^2$ according to Γ_3; and xy according to Γ_4. But what about x and y? We can hope that they may transform according to Γ_5 since the represent ation they generate is 2-dimensional. However it may be reducible. A method that we shall use in this case is the explicit construction of the representation

$$\hat{P}_R(x,y) = (x,y)\left(\Gamma(R)\right)$$

and from it we get the construction of the characters. Thus

$$\hat{P}_{a_x}(x,y) = (x,-y) = (x,y)\left(\Gamma(a_x)\right)$$

Hence

$$\Gamma(a_x) = \begin{pmatrix} 1 & 0 \\ 0 & -1 \end{pmatrix}$$

Similarly,

$$\Gamma(e) = \begin{pmatrix} 1 & 0 \\ 0 & 1 \end{pmatrix} \quad \Gamma(a_y) = \begin{pmatrix} -1 & 0 \\ 0 & 1 \end{pmatrix} \quad \Gamma(b) = \begin{pmatrix} -1 & 0 \\ 0 & -1 \end{pmatrix}$$

$$\Gamma(c_1) = \begin{pmatrix} 0 & -1 \\ 1 & 0 \end{pmatrix} \quad \Gamma(c_2) = \begin{pmatrix} 0 & 1 \\ -1 & 0 \end{pmatrix} \quad \Gamma(d_1) = \begin{pmatrix} 0 & 1 \\ 1 & 0 \end{pmatrix} \quad \Gamma(d_2) = \begin{pmatrix} 0 & -1 \\ -1 & 0 \end{pmatrix}$$

Taking the traces, we indeed find that (x,y) transforms according to Γ_5, as does (yz,xz), although the representation it generates is related to the one generated by (x,y) by an equivalence transformation.

Example D: The next example is the D_3 or 32 group discussed many times before. It is the group of the equilateral triangle.

	E	$2C_3$	$3C_2$
A_1	1	1	1
A_2	1	1	-1
E	2	-1	0

	x	y	z	x^2	etc.
E	x	y	z	x^2	
C_3	$\frac{1}{2}(-x+\sqrt{3}y)$	$\frac{1}{2}(-y-\sqrt{3}x)$	z	$\frac{1}{4}(x^2+3y^2-2\sqrt{3}xy)$	
C_3^{-1}	$\frac{1}{2}(-x-\sqrt{3}y)$	$\frac{1}{2}(-y+\sqrt{3}x)$	z	$\frac{1}{4}(x^2+3y^2+2\sqrt{3}xy)$	
$C_{2(1)}$	$-x$	y	$-z$	x^2	
$C_{2(2)}$	$\frac{1}{2}(x-\sqrt{3}y)$	$\frac{1}{2}(-y-\sqrt{3}x)$	$-z$	$\frac{1}{4}(x^2+3y^2-2\sqrt{3}xy)$	
$C_{2(3)}$	$\frac{1}{2}(x+\sqrt{3}y)$	$\frac{1}{2}(-y+\sqrt{3}x)$	$-z$	$\frac{1}{4}(x^2+3y^2+2\sqrt{3}xy)$	

Evidently z transforms as A_2, but the rest is not clear.
Once again we might work out explicitly the representations
generated by (x,y) and take traces. However we may save
ourselves this labor by doing the job once and for all in
general language and deriving a simple rule which enables
us to read the traces directly from the transformation
table. Thus if R is any element of G, then

$$\hat{P}_R(x,y) = (\alpha x+\beta y, \ \gamma x+\delta y) = (x,y) \ \begin{pmatrix}\Gamma(R)\end{pmatrix}$$

Therefore $\Gamma(R) = \begin{pmatrix}\alpha & \gamma \\ \beta & \delta\end{pmatrix}$ and $\chi(R) = \alpha + \delta$. Hence this rule:
Suppose that under a given operation $\hat{P}_R \phi_1 \rightarrow \alpha_1\phi_1+\alpha_2\phi_2+\ldots+\alpha_n\phi_n$
$\phi_2 \rightarrow \beta_1\phi_1+\beta_2\phi_2+\ldots+\beta_n\phi_n; \ldots; \phi_n \rightarrow \xi_1\phi_1+\xi_2\phi_2+\ldots+\xi_n\phi_n$. Then we
find the character $\chi(\mathbb{R})$ in this representation by adding
the coefficient ϕ_1 in the transformation of ϕ_1 to the
coefficient of ϕ_2 in the transformation of ϕ_2 and continue
the sum through the coefficient of ϕ_n in the transformation
of ϕ_n. That is, $\chi(R) = \alpha_1 + \beta_2 + \ldots + \xi_n$.
Thus to find $\chi(C_3)$ in the representation generated by (x,y), we add the coefficient of x in the transformation of x, $-\frac{1}{2}$, to that of y in the transformation of y,
$-\frac{1}{2}$, and we get -1. The list of characters in order of ele-

ments can now be read off at once: $(2, -1, -1, 0, 0, 0)$.
Thus we conclude that (x,y) transforms as E.

Finally we turn to the transformation of x^2. Rather than trying to guess at the components of x^2 which transform according to the irreducible representations of G, we project them out by means of the projection operators defined on page 57. For example, $\hat{\rho}^{(A_1)}x^2 = \frac{1}{6}\sum_R \chi^{(A_1)*}(R)\hat{P}_R x^2$. We have already calculated $\hat{P}_R x^2$ for all R in G; so we need only weight each result by the corresponding character and sum.

$$\hat{\rho}^{(A_1)}x^2 = \frac{1}{6}\{x^2+\tfrac{1}{4}(x^2+3y^2-2\sqrt{3}xy)+\tfrac{1}{4}(x^2+3y^2+2\sqrt{3}xy)+$$
$$x^2+\tfrac{1}{4}(x^2+3y^2-2\sqrt{3}xy)+\tfrac{1}{4}(x^2+3y^2+2\sqrt{3}xy)\}$$

$$= \frac{1}{6}\{3x^2+3y^2\} = \tfrac{1}{2}(x^2+y^2)$$

$$\hat{\rho}^{(A_2)}x^2 = \frac{1}{6}\{x^2+\tfrac{1}{4}(x^2+3y^2-2\sqrt{3}xy)+\tfrac{1}{4}(x^2+3y^2+2\sqrt{3}xy)-$$
$$x^2-\tfrac{1}{4}(x^2+3y^2-2\sqrt{3}xy)-\tfrac{1}{4}(x^2+3y^2+2\sqrt{3}xy)\}$$

$$= 0$$

$$\hat{\rho}^{(E)}x^2 = \frac{2}{6}\{2x^2-\tfrac{1}{4}(x^2+3y^2-2\sqrt{3}xy)-\tfrac{1}{4}(x^2+3y^2+2\sqrt{3}xy)+0+0+0\}$$

$$= \frac{1}{6}(3x^2-3y^2) = \tfrac{1}{2}(x^2-y^2)$$

Thus $x^2 = \tfrac{1}{2}(x^2+y^2) + \tfrac{1}{2}(x^2-y^2)$ and $x^2 + y^2$ transforms as A_1, while $x^2 - y^2$ transforms, along with some other quadratic form, as E. The other function is xy, as may be found by operating with $\hat{P}_{C_3}$ on $(x^2 - y^2)$ to get $\tfrac{1}{4}(x^2-y^2)-2\sqrt{3}xy$.

In this way we can show that (x,y), (x^2-y^2,xy), (xz,yz) all transform according to E; z according to A_2; z^2, x^2+y^2 according to A_1.

Example E: For a final example we choose the 12th order group of the tetrahedron introduced on page 14 and called T or 23. It is convenient to choose the 2-fold axes along the x-, y-, and z-axes so that the tetrahedron is inscribed inside a cube with its center at the origin.

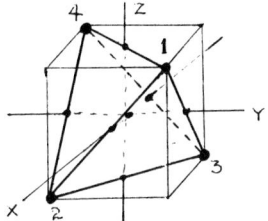

We write the character table and derive the transformation table.

	$\mathcal{C}_e$	$4\mathcal{C}_a$	$4\mathcal{C}_b$	$3\mathcal{C}_c$
A	1	1	1	1
E	1	ω	ω^2	1
	1	ω^2	ω	1
T	3	0	0	-1

	e	a_1	a_2	a_3	a_4	b_1	b_2	b_3	b_4	c_1	c_2	c_3
x	x	y	-y	-y	y	z	z	-z	-z	-x	-x	x
y	y	z	z	-z	-z	x	-x	-x	x	-y	y	-y
z	z	x	-x	-x	-x	y	y	-y	y	z	-z	-z
χ	3	0	0	0	0	0	0	0	0	-1	-1	-1
x^2	x^2	y^2	y^2	y^2	y^2	z^2	z^2	z^2	z^2	x^2	x^2	x^2
etc.												

The characters χ of the representation generated by (x,y,z) are calculated using the preceding rule. By using the projection operators on x^2, we find that it has the following three components $x^2+y^2+z^2$ transforming as A, and the two components $z^2+\omega^2x^2+\omega y^2$ and $z^2+\omega x^2+\omega^2y^2$, each of which transforms according to one of the representations of E. (According to the symmetry of T, these two functions should have different symmetries, and if they were eigenfunctions, they should be, if at all, only accidentally degenerate. In fact, owing to time reversal symmetry, they are always degenerate. Since the time reversal operator is not unitary - it is anti-unitary - it cannot be put into the symmetry group of the Schrödinger equation in the usual way. Nonetheless, its effect can be taken into account in a simple way, as we shall see later. The essential result is that eigenfunctions which transform according to representations which are complex conjugates of one another are necessarily degenerate.)

Since $z^2+\omega^2x^2+\omega y^2$ and $z^2+\omega x^2+\omega^2 y^2$ are degenerate, it is convenient to form the sum and difference linear combinations. This gives $2z^2-x^2-y^2$ and x^2-y^2. These functions are also orthogonal.

IV

Direct Product Theory

In the following sections we shall define the direct product of matrices and of groups, and we shall inquire into the representations of product groups.

1. THE DIRECT PRODUCT OF MATRICES

Let A be a matrix of elements a_{ij}, $i = 1\ldots\ell_{ar}$, $j = 1\ldots \ell_{ac}$, and let B be a matrix of elements b_{km}, $k = 1 \ldots\ell_{br}$, $m = 1\ldots\ell_{bc}$.

Definition: The matrix C is called the direct product of A and B, symbolized by writing $C \equiv A \times B$, if its elements are all of the $\ell_{ar}\,\ell_{ac}\,\ell_{br}\,\ell_{bc}$ products of each element of A with every element of B. A convenient labeling of the elements of C is defined by the equation $c_{ik,jm} = a_{ij}b_{km}$. The direct product is a commutative operation since the same array of elements is produced in whatever order all products of the elements of A and B are taken.

We are accustomed to think of matrices as two-dimensional arrays of numbers, although this is by no means necessary. We could, if we liked, string all of the elements out in a line. The important point is that when matrix multiplication is defined, only specific elements of one matrix multiply specific elements of the other, and so it is convenient to order the elements in the same way in every matrix and to choose a way which is convenient for matrix multiplication-- hence the rectangular array.

Before we decide on a convenient arrangement of elements of a direct product matrix we need a definition of matrix multiplication between direct product matrices. As ordinary matrix multiplication was defined so that matrices might represent transformations applied succesively in ordinary vector spaces, so matrix multiplication between product matrices is defined so that they may represent

succesive transformations applied in a "product space".
A product space is formed out of two ordinary spaces in
such a way that a transformation in the product space is a
combination of transformations carried out separately in
each of the ordinary spaces such that the rule for successive transformations is obeyed in each of the ordinary
spaces separately. Thus if A and A' represent operations
in space $\underline{a}$, and B and B' in space $\underline{b}$, then by ordinary
matrix multiplication AA' ≡ A'' is a matrix representing
the successive operations A', then A, in space $\underline{a}$, and similarly BB' ≡ B'' in space $\underline{b}$. Now if C ≡ A x B and C' ≡ A' x
B' are product matrices representing operations in the
product space, then we must define matrix multiplication
in this space so that CC' ≡ A'' x B'' = AA' x BB'. Or, in
terms of the matrix elements and with the labeling intro-
duced earlier, $(cc')_{ik,jm} = \sum_{pq} a_{ip} a'_{pj} b_{kq} b'_{qm} = \sum_{pq} c_{ik,pq} c'_{pq,jm}$.

Thus if we think of each distinct pair (i,k) as
labeling a <u>row</u> of a rectangular array $(c_{ik,pq})$ and each
pair (p,q) as labeling a column, then the rule for multiplying product matrices is exactly the same as that for
multiplying ordinary matrices.

As far as actually writing out the product matrices,
there is still the matter of ordering the pairs (i,k) into
some specific sequence of rows and columns. One convention
frequently followed is to set i = 1 and let k = 1, 2...ℓ_{br};
then set i = 2 and again give k its sequence of values, and
so continue until all i's have been exhausted. We proceed
in the same fashion with the pairs labeling the columns.
This convention is equivalent to the following: The matrix
of C ≡ A x B can be constructed by writing the matrix A
in the usual rectangular array and writing for each element
a_{ij} the product of a_{ij} with the entire rectangular matrix B.

$$C = A \times B = \begin{pmatrix} a_{11}B & a_{12}B \cdots \\ a_{21}B & a_{22}B \cdots \\ \vdots \end{pmatrix}$$

Thus the first row of C is $(a_{11}b_{11}, a_{11}b_{12} \cdots a_{11}b_{1\ell c}, a_{12}b_{11}$
... etc.) A different convention exchanges the a's and
b's. Actually we shall have little need of writing out
any matrices explicitly.

2. DIRECT PRODUCT OF GROUPS

Definition: If there are two groups G_a with elements a_i, i = 1...h_a, and G_b with elements b_j, j = 1 ...
h_b such that $a_i b_j = b_j a_i$ for all i and j, then the direct
product group $G_a \times G_b$ is defined to be the set of all distinct elements $a_i b_j$ produced for all i and j. If apart
from the identity G_a and G_b have no element in common,
then the order of the direct product group will be $h_a h_b$.

We show that the new set of elements satisfy the
group postulates.

1. Closure. $a_i b_j a_k b_\ell = a_i a_k b_j b_\ell = a_r b_s$. Since a_r
is in G_a and b_s is in G_b, $a_r b_s$ is, by definition, in $G_a \times$
G_b.

2. Unit element is $e_a e_b$.
3. Inverse of $a_i b_j$ is $a_i^{-1} b_j^{-1}$.
4. Associativity is obvious.

Theorem: The classes of the direct product group
are given by the direct products of the classes of the
original groups.

Proof: Let us label the elements of the product
group $c_{ij} = a_i b_j$. Now we seek the class which includes,
for a given i and j, the element c_{ij}. By definition of
classes, we produce the class of c_{ij}, which we call $\mathbf{C}_{ij}$,
by forming the complex of elements $\{c_{rs}^{-1} c_{ij} c_{rs}\}$ for all r
and s. Thus

$$\mathbf{C}_{ij} \equiv \{c_{rs}^{-1} c_{ij} c_{rs}\} , \text{ all r, s}$$
$$= \{a_r^{-1} b_s^{-1} a_i b_j a_r b_s\}$$
$$= \{a_r^{-1} a_i a_r b_s^{-1} b_j b_s\}$$
$$= \{a_r^{-1} a_i a_r\}\{b_s^{-1} b_j b_s\}$$

$$= c_i^a c_j^b \qquad Q.E.D.$$

Examples: (1) Unitary transformations in spin space and in coordinate space commute with one another and form groups in each space. (2) All permutations of particle coordinates commute with spatial relations. (3) The inversion commutes with all rotations.

Theorem: The direct products of the matrices representing G_a and G_b are representations of $G_a \times G_b$ under the matrix multiplication rule for direct product matrices.

Proof: The set of matrices $\{\Gamma(a_i)\}$ is a representation of G_a if $a_i a_j = a_k$ implies that, under matrix multiplication, $\Gamma_a(a_i)\Gamma_a(a_j) = \Gamma_a(a_k)$. Similarly for $\{\Gamma_b(b_i)\}$. As before we define $c_{ij} = a_i b_j$ to be an element of the product group. Now let us suppose that $c_{ij} c_{nm} = c_{rs}$ and discover what this implies. If $c_{ij} c_{nm} = c_{rs}$, then $a_i b_j a_n b_m = a_r b_s$ and $a_i a_n b_j b_m = a_r b_s$. But this can only be true if $a_i a_n = a_r$ and $b_j b_m = b_s$. And this implies that $\Gamma_a(a_i)\Gamma_a(a_n) = \Gamma_a(a_r)$ and that $\Gamma_b(b_j)\Gamma_b(b_m) = \Gamma_b(b_s)$. But this in turn implies that a specific direct product matrix $\Gamma_{cij} \equiv \Gamma_a(a_i) \times \Gamma_b(b_j)$ satisfies the equation under matrix multiplication for product matrices $\Gamma_{cij}\Gamma_{cnm} = \Gamma_{crs}$. Hence the new matrices do represent the direct product group, and we may write $\Gamma_{cij} = \Gamma_c(c_{ij})$.

Theorem: If $\Gamma_a^{(i)}(a_r)$ and $\Gamma_b^{(j)}(b_s)$ are irreducible representations of G_a and G_b, then $\Gamma_a^{(i)}(a_r) \times \Gamma_b^{(j)}(b_s) \equiv \Gamma_c^{(ij)}(c_{rs})$ is an irreducible representation of $G_a \times G_b \equiv G_c$.

Proof: Let M_c be a matrix which commutes with every matrix of $\{\Gamma_c^{(ij)}(c_{rs})\}$, that is, for all r,s. Evidently M_c must be of the correct dimensions to be written in the form $M_c = M_a \times M_b$ where M_a and M_b have the same dimensions as $\Gamma_a^{(i)}$ and $\Gamma_b^{(j)}$, respectively. Hence the commutativity of M with $\Gamma_c^{(ij)}(c_{rs})$ implies the commutativity of M_a and M_b with $\Gamma_a^{(i)}(a_r)$ and $\Gamma_b^{(j)}(b_s)$, respectively, for all r and s. But by Schur's Lemma, $M_a = k_a I_a$ and $M_b = k_b I_b$. Hence $M = M_a \times M_b = k_a k_b I_a \times I_b = k_{ab} I_c$. Thus no non-constant commuting matrix M_c exists, and so, by Schur's Lemma, $\Gamma_c^{(ij)}$ is irreducible.

Theorem: There are no other irreducible representations of $G_a \times G_b$.

Proof: By Theorem IX, page 45, $\sum_i \ell_{ai}^2 = h_a$, $\sum_j \ell_{bj}^2 = h_b$, and $\sum_{ij} \ell_{cij}^2 = h_c$. But $h_c = h_a h_b$. Now let h be the number defined by $h \equiv \sum_{ij} \ell_{cij}^2$, where the sum is over all irreducible representations generated by direct products. For them $\ell_{cij} = \ell_{ai} \ell_{bj}$. Thus $h = \sum_{ij} \ell_{ai}^2 \ell_{bj}^2 = \sum_i \ell_{ai}^2 \sum_j \ell_{bj}^2 = h_a h_b = h_c$, and there can be no more irreducible representations.

Application to construction of character tables. Suppose that to the group D_3 or 32 which consists only of rotations, and whose character table we have derived on page 37, we add inversion symmetry. The group with this symmetry is called D_{3d} or $\bar{3}m$ and may be derived by a direct product of D_3 with the inversion group $\{e,i\}$. Now $\{e,i\}$ is a second order cyclic group whose character table we have derived several times. Now since the character of a representation is a trace of the matrix, we must evidently inquire into the trace of direct product matrices. The row of such a matrix $\Gamma_{ij,km}$ is labeled by the pair of numbers (i,j) and the columns by (k,m). To form the trace we must add all elements whose row label and column label are identical. That is, i must equal k and j,m. Thus

$$\chi(a_i b_j) = \sum_{\mu\nu} \Gamma(a_i b_j)_{\mu\nu,\mu\nu} = \sum_{\mu\nu} \Gamma_a(a_i)_{\mu\mu} \Gamma_b(b_j)_{\nu\nu} = \chi_a(a_i) \chi_b(b_j).$$

So we get the character table quite trivially from the D_3 table and the second order group table. The ordering of classes is arbitrary but is chosen to display the fact that D_{3d} is a direct product group.

	E	$2C_3$	$3C_2$	i	$2iC_3$	$3iC_2$
	1	1	1	1	1	1
	1	1	-1	1	1	-1
	2	-1	0	2	-1	0
	1	1	1	-1	-1	-1
	1	1	-1	-1	-1	1
	2	-1	0	-2	1	0

Incidentally, if we think of the character table as a
square matrix, then the character table of a direct product
group is the direct product of the character tables of
the component groups.

3. DIRECT PRODUCT REPRESENTATIONS OF A GROUP WHICH
   ITSELF MAY NOT BE A DIRECT PRODUCT GROUP

Suppose we are interested in the transformation properties of a function which can be written as the product
of two simpler functions whose transformation properties
we already know. Our purpose here is to infer from them
some information about the product function.

Definition: If the set of vectors $\{\phi_\mu\} \equiv \underline{\phi}$ span a
subspace V_s, and if the set $\{\psi_\lambda\} = \underline{\psi}$ span a subspace $V_{s'}$,
then the set $\{\phi_\mu \psi_\lambda\} \equiv \underline{\phi} \times \underline{\psi}$ is said to span a product
space $V = V_s \times V_{s'}$. If the set $\underline{\phi}$ and $\underline{\psi}$ are basis vectors in
their respective subspaces, then $\underline{\phi} \times \underline{\psi}$ are basis vectors in
V. (Note that V_s and $V_{s'}$ may be identical, as may $\underline{\phi}$ and $\underline{\psi}$.)

Theorem: If V_s and $V_{s'}$ are invariant subspaces
under the operations of G, then so is $V_s \times V_{s'}$.

Proof: According to the definition of an invariant
subspace, $\hat{P}_R \underline{\phi} = \underline{\phi} \Gamma_1(R)$, $\hat{P}_R \underline{\psi} = \underline{\psi} \Gamma_2(R)$. But $\hat{P}_R(\underline{\psi} \times \underline{\phi}) = (\hat{P}_R \underline{\psi}) \times$
$(\hat{P}_R \underline{\phi}) = (\underline{\phi} \Gamma_1(R)) \times (\underline{\psi} \Gamma_2(R)) = (\underline{\phi} \times \underline{\psi})(\Gamma_1(R) \times \Gamma_2(R))$. Since
this is true for all R, the theorem is proved. Theorem (i),
page 53, tells us that $\Gamma_1(R) \times \Gamma_2(R)$ is therefore a representation of G. Thus we may say that the set of vectors
$\underline{\phi} \times \underline{\psi}$ transforms according to $\Gamma_1 \times \Gamma_2$. But $\Gamma_1 \times \Gamma_2$ need
not be irreducible even if Γ_1 and Γ_2 are. If $\Gamma_1 \times \Gamma_2$ is
reducible, then it may be decomposed, or shown in Theorem
VIII, page 42, into its irreducible parts.

Examples: To show the utility of this result, we
return to the example of the D_3 group discussed on pages
64-66. There we found that

(x,y) transforms according to representation E

   z   transforms according to representation A_2

(i) How does the product space

$$(xz, yz) = (x, y) \times (z)$$

transform? We multiply character of E by character of A_2, element by element, and obtain

$$(2, -1, 0) \times (1, 1, -1) = (2, -1, 0)$$

and since the trace of a direct product of two matrices equals the product of the traces,

$$A \times B = C$$

$$C_{ij, k\ell} = A_{ik} B_{j\ell}$$

$$TrC = \sum_{ij} C_{ij, ij} = \sum_{ij} A_{ii} B_{jj} = [Tr\ A].[Tr\ B],$$

the character of the direct product of two representations is the product of the characters, and so

$$E \times A_2 = E.$$

Consequently (xz, yz) transforms according to E, as seen before.

(ii) How does

$$(x, y) \times (x, y) = (x^2, xy, yx, y^2)$$

transform? Multiplying characters again, we obtain

$$(2, -1, 0) \cdot (2, -1, 0) = (4, 1, 0)$$

and it is evident that the invariant subspace (x^2, xy, yx, y^2) is reducible.

$$(4, 1, 0) = (1, 1, 1) + (1, 1, -1) + (2, -1, 0),$$

or $$E \times E = A_1 + A_2 + E.$$

We have already found the subspaces $(x^2 + y^2)$ and $(x^2 - y^2, xy)$ which transform according to A_1 and E, respectively, and since only three functions are linearly independent, the contribution to the A_2 subspace should vanish. However if we look at the direct product

$$(x_1, y_1) \times (x_2, y_2) = (x_1 x_2, x_1 y_2, y_1 x_2, y_1 y_2)$$

where 1 and 2 refer to different "particles", then

$$(x_1 x_2 + y_1 y_2) \qquad \text{transforms according to } A_1$$
$$(x_1 x_2 - y_1 y_2, x_1 y_2 + y_1 x_2) \qquad \text{transform according to } E$$
$$(x_1 y_2 - x_2 y_1) \qquad \text{transforms according to } A_2.$$

V

Applications to Crystal and Molecular Symmetry

We first discuss the properties of the groups which may arise in our study.

I. <u>The group of all rotations about a given axis</u>. This is a continuous abelian group, and, although we have only proved it for finite abelian groups, it is true here also that the representations are all one-dimensional. We label the elements $C(\varphi)$ where φ is the angle of rotation, and the $n\underline{th}$ representation is labeled $\Gamma^{(n)}(\varphi)$. Now since $C(\varphi_1)\ C(\varphi_2) = C(\varphi_1+\varphi_2), \Gamma^{(n)}(\varphi_1)\Gamma^{(n)}(\varphi_2) = \Gamma^{(n)}(\varphi_1+\varphi_2)$. The most general continuous function of φ which satisfies this equation is $\Gamma^{(n)}(\varphi) = e^{k_n\varphi}$. Now, in addition, we must satisfy the condition that $\Gamma^{(n)}(2m\pi) = \Gamma^{(n)}(0) = 1$, since $C(0) = C(2m\pi)$ is the identity element. Hence, $e^{2m\pi k_n} = 1$. Evidently this will be so only if $k_n m = i$ x (integer). By convention we label the representation with the value of the integer. That is, $k_n = i\frac{n}{m}$ and

$$\Gamma^{(n)}(\varphi) = e^{in\varphi}$$

II. <u>The group of translations by an amount a</u>. This is a cyclic group with a generating element T_a such that

$$T_a\mathbf{x} = \mathbf{x}-a \text{ and } T_{na}\mathbf{x} = (T_a)^n\mathbf{x} = \mathbf{x}-na .$$

We make the group finite by imposing periodic boundary conditions, that is, by imposing

$$T_{Na} \text{ x} = \text{x}-Na = \text{x} .$$

T_{Na} is the identity element. The irreducible representa-

tions, as we have seen for cyclic groups (page 40) in
general, are obtained by assigning each one of the N N$\underline{\text{th}}$
roots of unity to the generating element T_a; we may then
label the representation

$$\Gamma^{(r)}(na) = e^{2\pi i rn/N} \qquad r = 0, 1, \ldots (N-1)$$

If we let Na = L, na = a_n, and define

$$k \equiv 2\pi r/L ,$$

then since k and r are in a one-to-one correspondence, we
may label the irreducible representations by k instead of r

$$\Gamma^k(a_n) = e^{ika_n}$$

Let us now find the properties of the functions which
transform according to Γ^k.

$$\hat{P}_{Ta} \, \Phi_k(x) = \Phi_k(x) \, \Gamma^k(a)$$

$$\Phi_k(x+a) = \Phi_k(x)e^{ika} .$$

Clearly every function of the form

$$\Phi_k = u_k(x) \, e^{ikx}$$

where

$$u_k(x+a) = u_k(a)$$

satisfies this equation. This is a proof of Bloch's
theorem. Physically we have proved that all the eigenfunctions of a Hamiltonian with translational symmetry can
be expressed in this so-called Bloch form.

III. Point groups of the crystal. These thirty-two
groups are composed of proper and improper rotations consistent with the translational symmetry of the crystal
lattice. The elementary operations may be thought of as
suitable products of proper rotations by angles $\frac{2\pi}{n}$ about
a specified axis, symbolized by C_n in the Schoenflies notation and by n in the international rotation; reflections,

symbolized by σ and m in the two systems; and <u>inversions</u>, symbolized by S_2 and $\overline{1}$. But there is some redundancy here, because there are actually only two species of rotation — the proper and the improper. Thus, given any two of the classes of elementary operations, the third may be expressed in terms of the other two. The Schoenflies notation emphasizes the proper rotations and reflections. The inversion is then reached by a two-fold rotation and a reflection through a plane perpendicular to the two-fold axis - hence the symbol S_2. The international notation emphasizes the proper rotations and the inversion. The reflection across a given plane is then reached by a two-fold rotation about an axis normal to the plane and the inversion. This operation is called $\overline{2}$ for this reason or m because it is a mirror operation.

Before enumerating the possible types of operations and the point groups, we prove the following theorem.

<u>Theorem</u>: The only possible proper rotations consistent with translational symmetry are C_n, where n = 1, 2, 3, 4, and 6.

<u>Definition</u>: A three-dimensional lattice of points is said to have translational symmetry if there exists some set of so-called primitive translations a_1, a_2, a_3 such that every point in the lattice may be reached from another one by a translation of the form

$$\underline{T}(t_1, t_2, t_3) = t_1\underline{a_1} + t_2\underline{a_2} + t_3\underline{a_3}$$

where t_1, t_2, and t_3 are all integers.

<u>Proof</u>: (1) First we show that if any rotational symmetry axis exists, then translational symmetry requires the existence of translation elements in planes perpendicular to the axis. The proof is simple. If C_n is an n-fold rotation about a given axis, the application of C_n to one translation vector $\underline{T}$ not parallel to the axis should yield another, different translation vector; that is

$$C_n \, \underline{T}(t_1, t_2, t_3) = \underline{T}(s_1, s_2, s_2).$$

But by definition of translation vectors, the difference

$$\underline{T}(t_1-s_1, t_2-s_2, t_3-s_3) = T(t_1, t_2, t_3) - T(s_1, s_2, s_3)$$

is another translation vector and by construction it is perpendicular to the n-fold axis. This can be easily seen if we realize that $\underline{T}(t_1-s_1, t_2-s_2, t_3-s_3)$ is a vector connecting the end point of the original vector to that of the rotated vector.

(2) To prove the theorem, let us now consider all vectors $\underline{T}(t_1, t_2, t_3)$ perpendicular to the n-fold axis. Of these we select one vector T_1 (other than the identity or zero translation) whose length $|\underline{T}_1| = T_1$ is shorter than or equal to any of the others. We apply to $\underline{T}_1$ the operations C_n, C_n^{2}, C_n^{3} ... corresponding to rotations by $2\pi/n$, $4\pi/n$, $6\pi/n$.., respectively, and impose the restriction that the only elements C_n to be retained are those for which the vectors

$$\underline{T}_{nm} = C_n^m \, \underline{T}_1$$

$$\underline{T}_{nm}^{\,1} = \underline{T}_{nm} - \underline{T}_1$$

$$\underline{T}_{nm}^{\,2} = \underline{T}_{nm} + \underline{T}_1 \ ,$$

are all translation elements of length either equal to or larger than T_1 or equal to zero (the identity).

Case (1): n = 1. In this case, C_1 is the identity and the conditions are satisfied.

Case (2): n = 2. In this case, $C_2^{2} = C_1$ and

$$\underline{T}_{21} = C_2 \, \underline{T}_1 = -\underline{T}_1 \ ,$$

Because of the existence of the inverse in the translation group, $-\underline{T}_1$ is a translation element and

$$|\underline{T}_{21}| = |-\underline{T}_1| = T_1$$

$$|\underline{T}_{21}^{\,1}| = |(-\underline{T}_1) - \underline{T}_1| = 2 \, T_1 > T_1$$

$$| \underline{T}_{21}^2 | = | (-\underline{T}_1) + \underline{T}_1 | = 0 .$$

The conditions are obviously satisfied.

Case (3): $n \geq 3$, $m < n$.

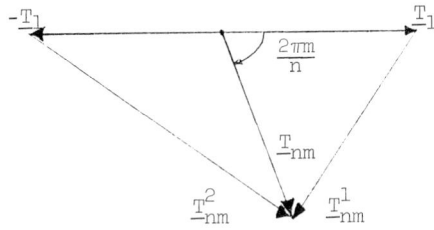

It is evident from the figure above that

$$| \underline{T}_{nm} | = | \underline{T}_1 | = T_1$$

$$| \underline{T}_{nm} \pm \underline{T}_1 | = T_1 [1 + 1 \pm 2 \cos \frac{2\pi m}{n}]^{1/2}$$

and consequently

$$| T_{nm}^1 | = 2 T_1 | \sin \frac{\pi m}{n} |$$

$$| T_{nm}^2 | = 2 T_1 | \cos \frac{\pi m}{n} | .$$

Since these two quantities should be either zero or else equal to or greater than T_1 , the only values of n which are compatible with the translation group are those which satisfy

$$\frac{1}{6} \leq \frac{m}{n} \leq \frac{1}{3} \quad \text{or} \quad \frac{m}{n} = \frac{1}{2} \quad \text{or} \quad \frac{2}{3} \leq \frac{m}{n} \leq \frac{5}{6}$$

for all values of m such that $0 < m < n$. It can be easily seen that in addition to $n = 1$ and $n = 2$, only $n = 3$, 4, and 6 satisfy the conditions. For $n = 5$, the conditions are satified by $m = 1$ and $m = 4$ but not by $m = 2$ or $m = 3$. For $n \geq 7$, the case $m = 1$ does not satisfy the criterion.

FUNDAMENTAL POINT GROUP OPERATIONS AND NOMENCLATURE

In both international and Schoenflies notations, the n-fold proper rotations are considered elementary operations. In the former, they are called n and in the latter, C_n. To get the remaining operations we form products of each n with i, the inversion operator, and label the result $\bar{n}$, in international notation. Alternatively, we form the product of each C_n with the operation σ_h, reflection in a plane perpendicular to the n-fold axis, and label the result S_n, except for $S_1 \equiv \sigma$. Evidently there is a one-to-one correspondence between the improper rotations, and this is shown in the table below. The order within the table is appropriate for the international notation.

Proper Rotations		Improper Rotations	
International	Schoenflies	International	Schoenflies
1	C_1	$\bar{1}$	S_2
2	C_2	$\bar{2} \equiv m$	σ
3	C_3	$\bar{3}$	S_6^{-1}
3_2	C_3^{-1}	$\bar{3}_2$	S_6
4	C_4	$\bar{4}$	S_4^{-1}
4_3	C_4^{-1}	$\bar{4}_3$	S_4
6	C_6	$\bar{6}$	S_3^{-1}
6_5	C_6^{-1}	$\bar{6}_5$	S_3

If the group has a principal axis, σ_h designates a mirror plane perpendicular to that axis, σ_v designates a mirror plane containing the axis, and σ_d designates a mirror plane containing the axis but diagonal to an already existing σ_v.

CONSTRUCTION AND ENUMERATION OF THIRTY-TWO POINT GROUPS

Since this list will proceed from simplest to most complex (in some sense), it is well to inquire into the simplest way of adding successive complications to already existing groups. Evidently the addition of a single new element, for example, inversion, which commutes with every element of the group, will double the size of the group in a particularly simple way. The following rules regarding commutativity will be helpful in enlarging simple groups.

(1) Inversion commutes with all operations.

(2) All rotations about same axis commute.

(3) Hence all rotations commute with a reflection across a plane perpendicular to the rotation axis.

(4) Two two-fold rotations about perpendicular axes commute.

(5) Two reflections in perpendicular planes commute.

The addition of one new element will imply the existence of others. These rules then apply.

(6) An n-fold axis (n>2) and a perpendicular two-fold axis imply additional two-fold axes.

(7) C_n (n≥2) and σ_v imply additional σ_v's.

(8) Any two of the following elements imply the third: σ_h, S_2, C_n (n even).

The Stereogram. The clearest way of picturing the symmetries of a group of rotations is to follow the motion of a representative point on the surface of a sphere as each element of the group is applied to the sphere. On the other hand, three-dimensional objects are hard to draw. Clearly a one-to-one mapping of the points on the sphere onto a plane is desirable. The stereogram gives that. For point group use, the stereogram is constructed as follows: project every point on the north hemisphere onto the plane PL, using straight-line projection through the south pole S. Use the north pole N as the projection point for the south hemisphere. Distinguish the upper hemisphere points by crosses and lower ones by circles.

The stereogram is clearly a one-to-one mapping, and it has the further property that points lying on a (not necessarily great) circle on the sphere map onto a circle on the plane, although the centers do not map onto one another. Thus the great circle cDe in the sphere of the figure maps onto cde in the stereogram.

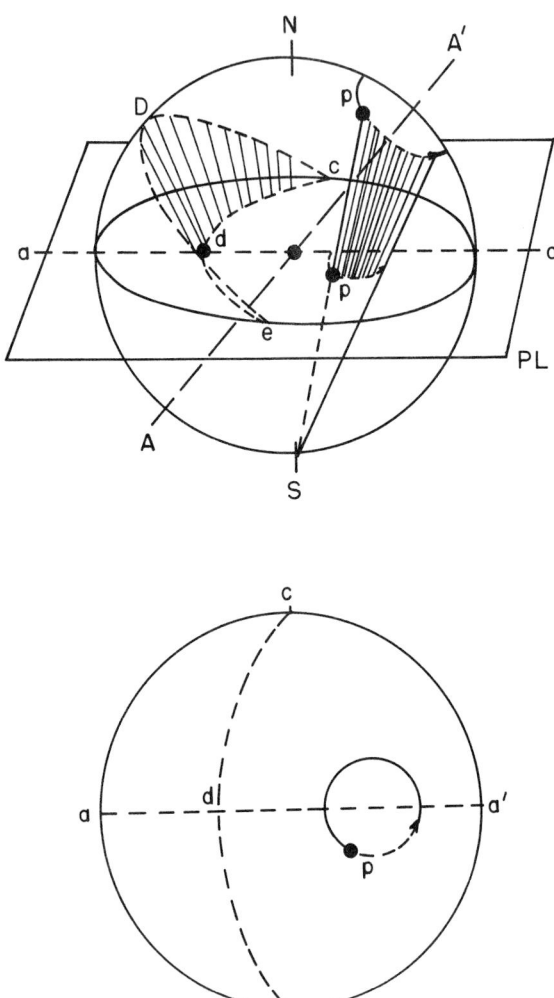

Where possible the polar axis of the stereogram coincides with the principal axis of symmetry. But in groups of higher symmetry this is not always possible, and all axes cannot be oriented either along the polar axis or in the projection plane. Such an axis might be AA' in the figure on page 84 . Thus it projects onto aa' (and its equator cDe goes to cde, as we have seen). Furthermore, a rotation about AA' will move the point P in a circle. Its projection must also move in a circle in the stereogram, as shown.

We shall also adhere to the following conventions in presenting stereograms of the thirty-two point groups:

(1) Symbols for 2-, 3-, 4-, and 6-fold rotation axes are $\bigcirc$, $\triangle$, $\square$, $\hexagon$.

(2) Solid lines represent mirror planes; all other lines are dashed. Thus axes and the outline of the stereogram will be shown as dashed lines unless they coincide with mirror planes.

(3) One particular general point and all points symmetric with it are represented by crosses and circles. (Evidently the number of equivalent points is equal to the order of the group).

(4) There seems to be no general agreement on how to represent the presence of inversion; therefore, we merely point out in which groups it is an element.

ENUMERATION OF THE THIRTY-TWO POINT GROUPS

We now present stereograms, Schoenflies symbols, short international symbols, and full international symbols where they differ from the short form.

The first 5 groups have a single n-fold axis.

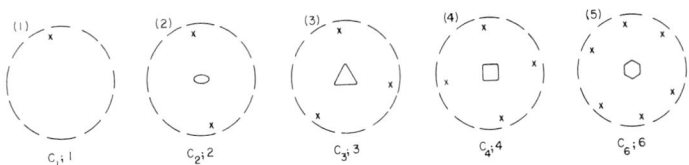

From these five we get five more by adding a horizontal mirror. Evidently C_{nh}, n even, have inversion symmetry.

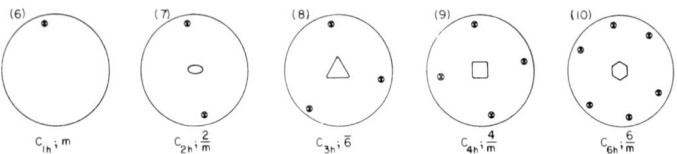

We generate four more groups by adding a vertical mirror plane instead of a horizontal mirror plane to the original five groups. Rule 7, page 83 , implies additional mirror planes. Since "vertical" and "horizontal" are not distinct for group C_1, we get no new group here.

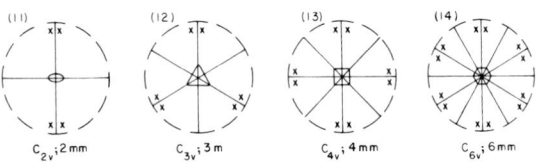

Before we add more n-fold axes, we can generate three more groups. The first five groups are cyclic groups generated by successive application of proper rotations about a given axis. Now we apply the improper rotations. But σ produces the group C_{1h}, and S_3 produces C_{3h}. The new groups are

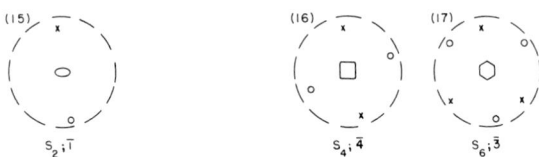

S_2 and S_6 have inversion symmetry. Next we add a two-fold axis perpendicular to the n-fold axes of the first five groups. Rule 6, page 83, implies additional two-fold axes. The new groups are

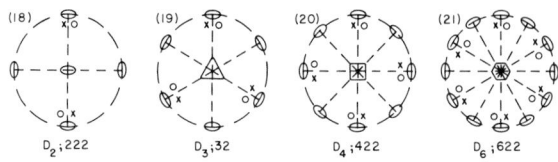

$D_2;222$ $D_3;32$ $D_4;422$ $D_6;622$

Next we add a two-fold axis perpendicular to the n-fold axes (and hence in the horizontal mirror plane) of the second five groups, which we have labeled C_{nh}, page 86. C_{1h} and two-fold axis is just C_{2v}. We get four new groups.

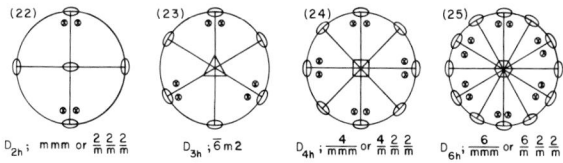

$D_{2h}; \; mmm \; or \; \frac{2}{m}\frac{2}{m}\frac{2}{m}$ $D_{3h}; \bar{6}m2$ $D_{4h}; \frac{4}{mmm} \; or \; \frac{4}{m}\frac{2}{m}\frac{2}{m}$ $D_{6h}; \frac{6}{mmm} \; or \; \frac{6}{m}\frac{2}{m}\frac{2}{m}$

The additional axes are implied by rule 6, and the additional mirror planes are implied by rule 7, page 83. (The D_{nh} groups may also be thought of as arising from the addition of a horizontal mirror plane to the groups D_h.) D_{nh}, n even, have inversion symmetry. Finally we add a two-fold axis to the groups labeled S_n. The result for n = 2 is the group C_{2h}. We get two new groups from S_4 and S_6.

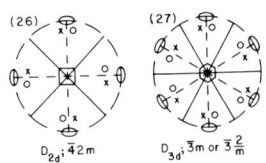

$D_{2d}; \bar{4}2m$ $D_{3d}; \bar{3}m \; or \; \bar{3}\frac{2}{m}$

D_{2d}, like S_4, has a true two-fold axis perpendicular to
the stereogram. Similarly D_{3d} and S_6 have three-fold axes.
D_{3d} has inversion symmetry.

This completes our list of simple groups in which
there is a principal axis with all other axes perpendicular
to it. There are five groups of higher symmetry and they
are all characterized by having no single principal axis
and by having a three-fold axis which lies equidistant from
three mutually perpendicular two- or four-fold axes.
(28) This is the group of <u>proper</u> rotations of the tetrahedron. We orient the tetrahedron as shown on page 67 and
take one of the co-ordinate axes, let us say the z-axis,
as the polar axis of the stereogram.

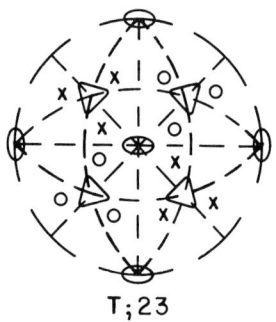

T;23

Even though the planes of mirror symmetry do not enter into
this particular group, it is conventional to display their
locations in the stereogram. This is convenient because
their intersections locate the ends of the three-fold axes
on the sphere from which the stereogram is made. To indicate that they are <u>not</u> mirror planes in T, we draw them
with dashed lines. The two which contain the z-axis appear
in the stereogram as straight lines which coincide with the
three-fold axes. The other four, which contain the x- and
y-axes, project as circles in the stereogram. Apart from

the three-fold axes, the other elements of T are the same
as the D_2 group; this is perhaps the easiest way to com-
prehend the T group.

(29) The full tetrahedral group contains, in addition to T,
all mirror planes. We symbolize the two-fold axes as in T,
but the fact is that they are axes of the S_4 type (a four-
fold rotation and a perpendicular mirror).

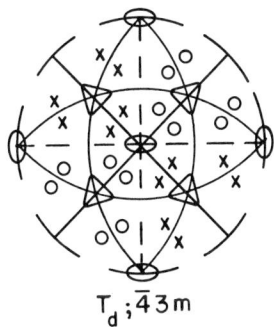

$$T_d; \bar{4}3m$$

(30) We get a different group from the tetrahedral group
T by adding inversion symmetry but not the mirror planes
of T_d. (The tetrahedron itself does not have this symmetry.)

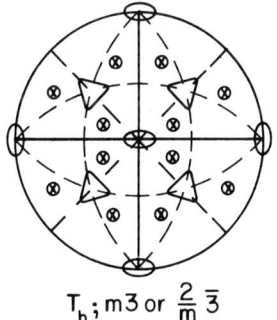

$$T_h; m3 \text{ or } \frac{2}{m}\bar{3}$$

(31) The <u>proper</u> rotation group of the octahedron (or of the cube) consists of mutually orthogonal four-fold axes. This is enough to imply all the other n-fold axes in the stereogram.

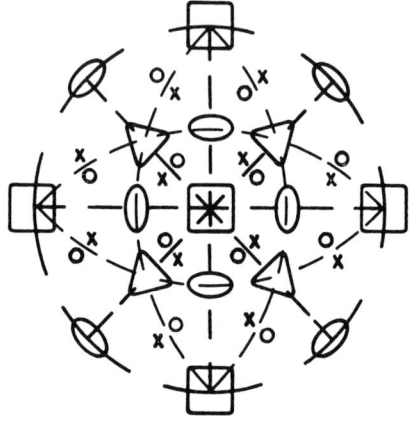

$$O\,;\,432$$

(32) The full octahedral group is obtained by adding inversion symmetry either to O or to T_d. The three-fold axes have additional symmetry of the S_6 type. Every axis contains a mirror plane.

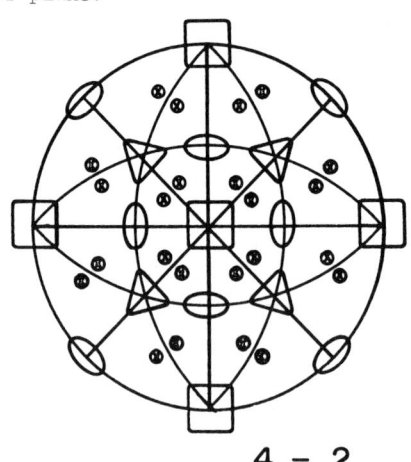

$$O_h\,;\,m3m \text{ or } \frac{4}{m}\,\bar{3}\,\frac{2}{m}$$

This completes the list. Perhaps we should emphasize at this point that finite groups of rotations, other than these thirty-two, exist; but none of them is consistent with lattice symmetry. The group of the dodecahedron, for example, contains a 5-fold axis and so must be disallowed. Since perfect crystals have a lattice structure, we shall only be interested in the crystallographic point groups when we study solids. It so happens that the symmetries of many free molecules are to be found among these same groups.

The following general statements are useful in finding representations of these point groups.

1. Groups numbered 1, 2, 3, 4, 5, 6, 8, 15, 16, and 17 are all cyclic. (These are the groups labeled n and $\overline{n}$ in the international notation.)

2. Groups numbered 7, 9, 10, 17, 22, 24, 25, 27, 30 and 32 are direct products of the inversion group with groups 2, 4, 5, 3, 11, 13, 14, 19, 28 and 31, respectively. (These direct product groups are labeled with one or more m's in the denominator in the international notation, except for number 17.)

We have enumerated these groups in a way which is appropriate for the Schoenflies notation, but there is a very convenient way of tabulating our results from the point of view of the international notation. First, we list in a row all the cyclic groups labeled n; second, the remaining cyclic groups labeled $\overline{n}$; third, omitting redundancies, the horizontal mirror groups labeled n/m; fourth, the distinct vertical mirror groups labeled nm; fifth, the distinct groups found by adding a vertical mirror to $\overline{n}$, labeled $\overline{n}$m; sixth, the distinct groups with two-fold axes perpendicular to n-fold axes, labeled n2; and finally the groups with horizontal and vertical mirrors and an n-fold axis, labeled $\frac{n}{mm}$. Then we make one more column which starts with the group labeled 23 and progresses by making the same changes on the 2 that we have made on the n in the successive rows. This arrangement of the groups is convenient because the columns then give the crystal system with which the groups therein are compatible.

The crystal systems are defined in terms of the properties of the group of translations. Given three linearly independent translations $\underline{a}$, $\underline{b}$, $\underline{c}$ (which are not necessarily the three primitive translations $\underline{a}_1$, $\underline{a}_2$, $\underline{a}_3$), the magnitude of the vectors - a, b, and c - , and the angles between them, - α $(\underline{b}, \underline{c})$, β $(\underline{c}, \underline{a})$, and γ $(\underline{a}, \underline{b})$ - determine the crystal system.

System	Properties
Cubic	$a = b = c$ $\qquad \alpha = \beta = \gamma = \frac{\pi}{2}$
Hexagonal	$a = b \neq c$ $\qquad \alpha = \beta = \frac{\pi}{2}$ $\gamma = \frac{2\pi}{3}$
Rhombohedral	$a = b = c$ $\qquad \alpha = \beta = \gamma < \frac{2\pi}{3} \neq \frac{\pi}{2}$ and $\neq \frac{\pi}{3}$
Tetragonal	$a = b \neq c$ $\qquad \alpha = \beta = \gamma = \frac{\pi}{2}$
Orthorhombic	$a \neq b \neq c$ $\qquad \alpha = \beta = \gamma = \frac{\pi}{2}$
Monoclinic	$a \neq b \neq c$ $\qquad \alpha = \gamma = \frac{\pi}{2} \neq \beta$
Triclinic	$a \neq b \neq c$ $\qquad \alpha \neq \beta \neq \gamma$

	Tri-clinic	Mono-clinic or Ortho-rombic	Rhombo-hedral	Tetra-gonal	Hexa-gonal	Cubic
n	1	2	3	4	6	23
$\bar{n}$	$\bar{1}$	$\bar{2} \equiv m$	$\bar{3}$	$\bar{4}$	$\bar{6}$	$23 \equiv m3$
$\frac{n}{m}$	$(=m)$	$\frac{2}{m}$	$(=\bar{6})$	$\frac{4}{m}$	$\frac{6}{m}$	$(=m3)$
nm	$(=m)$	$2mm$	$3m$	$4mm$	$6mm$	$(=m3)$
$\bar{n}m$	$(=\frac{2}{m})$	$(=2mm)$	$\bar{3}m$	$\bar{4}2m$	$\bar{6}m2$	$\bar{4}3m$
$n2$	$(=2)$	222	32	422	622	432
$\frac{n}{mm}$	$(=2mm)$	mmm	$(=\bar{6}m2)$	$\frac{4}{mmm}$	$\frac{6}{mmm}$	$m3m$

Character tables of the thirty-two point groups are widely available in standard textbooks of crystallography and group theory; so we shall not quote them here. When we need a particular table we shall reproduce it at that time.

PHYSICAL APPLICATIONS

By applying the theory of the symmetry of the Hamiltonian and the theory of groups to a model problem, we propose to answer the question, "How do the degenerate energy levels of a free atom split when the atom is placed in a crystal lattice of a given symmetry?"

At the start, to make the problem as simple as possible, we shall assume the atom to have a single, spinless electron with a series of hydrogen-like levels. In the free atom we find that all states of the same n and ℓ quantum numbers are $(2\ell+1)$-fold degenerate.

The theory of symmetry suggests that this degeneracy is a reflection of some symmetry property of the Hamiltonian of the free atom; and this is true. This Hamiltonian has the symmetry of the full rotation group, a continuous group which we shall study in detail later. For the present we may make plausible some results rigorously proved there.

First we recall from elementary quantum mechanics that the set of functions called the spherical harmonics, and symbolized by $Y_\ell^m(\theta,\varphi)$, is a complete set for expressing arbitrary angular dependence. In fact, in the language of vector algebra, the set $\{Y_\ell^m\}$ for a particular ℓ is the basis of a $(2\ell+1)$-dimensional irreducible subspace invariant under the operations of the full rotation group. The invariance of the subspace implies the existence of coefficients $D^\ell(R)_{m'm}$ for all R, such that

$$\hat{P}_R \ Y_\ell^m = \sum_{m'=-\ell}^{\ell} Y_\ell^{m'} \ D^\ell(R)_{m'm} \ .$$

Theorem (i), page 53 , says that $D^\ell(R)$ is a representation of the group. The irreducibility of the subspace and theorem (v), page 54 , say that the representation is irreducible. Now since the full rotation group is the group of the Schrödinger equation for the free atom, it follows that the degenerate eigenfunctions of a given energy must transform according to an irreducible representation of this group. If the degeneracy is normal, <u>all</u> eigenfunctions of the level transform according to the <u>same</u> irreducible representation. In this case, the degeneracy of the level is $(2\ell+1)$ if the eigenfunctions transform according to the ℓth representation. Later we show the connection of ℓ with angular momentum.

All proper rotations through the same angle α about any axis belong to the same class of the full rotation group. This is so because it is always true that two rotations through the same angle are in the same class if there is a rotation in the group which takes one of the axes into the other which is certainly true of the full group.

To find the character table of this group we choose the axis to be the z-axis and recall that $Y_\ell^m = C_{\ell m} P_\ell^m(\cos\theta)e^{im\varphi}$. Hence

$$\hat{P}_{(\alpha,z)} \, Y_\ell^m = e^{-im\alpha} \, Y_\ell^m \; ;$$

so

$$D^\ell(\alpha,z)_{m'm} = \delta_{m'm} \, e^{-im'\alpha} \; .$$

Since elements of the same class have the same character,

$$\chi^\ell(\alpha) \equiv \chi(\alpha,z) = \sum_{m=-\ell}^{\ell} e^{-im\alpha} = e^{-i\ell\alpha} \sum_{m=0}^{2\ell} e^{+im\alpha} =$$

$$e^{-i\ell\alpha} \frac{1-e^{i(2\ell+1)\alpha}}{1-e^{i\alpha}} = e^{-i(\ell+\frac{1}{2})\alpha} \frac{1-e^{i(2\ell+1)\alpha}}{e^{-i\alpha/2}-e^{i\alpha/2}} = \frac{\sin(\ell+\frac{1}{2})\alpha}{\sin\frac{\alpha}{2}}$$

$\ell = 0, 1, 2 \cdot \cdot \cdot \cdot$

This important result is all that we need here.

Now we may turn to the solution of our problem. In order to be more specific we first place our atom in a crystal potential with the cubic symmetry of the O group and seek the splittings permitted by the loss of symmetry. Next we stress the cubic lattice along the (111) direction of the three-fold axis, bringing the symmetry down to D_3. Then we relax it and apply stress in the (110) direction, bringing the symmetry to D_2. We need character tables for all these groups.

Group	Repr.		$E(\alpha=0)$	$C_2(\alpha=\pi)$		$C_3(\alpha=\frac{2\pi}{3})$	$C_4(\alpha=\frac{\pi}{2})$
D	D^0 (s)		1	1		1	1
	D^1 (p)		3	-1		0	1
	D^2 (d)		5	1		-1	-1
	D^3 (f)		7	-1		1	-1
	D^4 (g)		9	1		0	1
			E	$3C_4^2$	$6C_2$	$8C_3$	$6C_4$
O	A_1 Γ_1		1	1	1	1	1
	A_2 Γ_2		1	1	-1	1	-1
	E Γ_3		2	2	0	-1	0
(432)	T_1 Γ_4		3	-1	-1	0	1
	T_2 Γ_5		3	-1	1	0	-1
			E		$3C_2$	$2C_3$	
D_3	A_1		1		1	1	
(32)	A_2		1		-1	1	
	E		2		0	-1	
			E	C_z	C_y	C_x	
D_2	A_1		1	1	1	1	
(222)	B_1		1	1	-1	-1	
	B_2		1	-1	1	-1	
	B_3		1	-1	-1	1	

There are, of course, an infinite number of representations and classes for D, but we indicate only a few. These tables are so arranged that classes of rotation by $\frac{2\pi}{n}$ fall in the same general column, which is appropriate for this problem.

Now let us consider the splitting of an f-level in the free atom when it is subjected to an external potential with cubic symmetry. The seven eigenfunctions will transform according to the irreducible representation D^3 of the full group; but in general they will not transform as an irreducible representation of O. Nevertheless, they will generate some representation of O because all operations of O are operations of D. Our task then is to reduce that representation into its irreducible components.

We do not have to know the eigenfunctions in order to do this, for we have the characters under (any) two-fold, three-fold, and so on, rotations; in particular we have the characters under those of D. Thus the characters generated by the operations of O on the seven f-functions are (7, -1, -1, 1, -1) in the same order as the table. Now we may reduce this by Theorem VIII, page 42. We then find that $D^3 \rightarrow A_2 + T_1 + T_2$. Thus the 7-fold level splits into two three-fold levels and one non-degenerate level - apart from accidental degeneracy.

Now let us remove further symmetry. If we strain the lattice uniformly along the (111) direction, then that three-fold axis remains but the others are destroyed. The six two-fold axes in the O group lie along the (011), (101), (110), and (01$\bar{1}$), (10$\bar{1}$) and (1$\bar{1}$0) directions. The last three axes are evidently perpendicular to (111); so they remain in the new group. Hence the new group is D_3 (or 32), the group of the equilateral triangle.

We have written its character table on page 95 in such a way as to show which elements of O are also in D_3. Once again we argue that since every element of D_3 is also in O we may deduce the characters of the (generally reducible) representations of D_3 generated by a subspace of functions transforming according to some irreducible representation of O. For example, the functions of O which transform according to A_2 generate a representation of D_3 whose characters are (1, -1, 1), in the order of the classes of D_3 shown in the table. Similarly T_1 of O $\rightarrow$ (3, -1, 0) in

D_3; and T_2 of $0 \rightarrow (3, 1, 0)$ in D_3. The first of these new representations is itself irreducible. The other two may be reduced to the representations $A_2 + E$ and $A_1 + E$ of D_3, respectively.

We represent these results symbolically by an energy level diagram, where each level is labeled by the name of the irreducible representation according to which its eigenfunctions transform. The degeneracy, which is the character of the identity element in that representation, is indicated in parentheses.

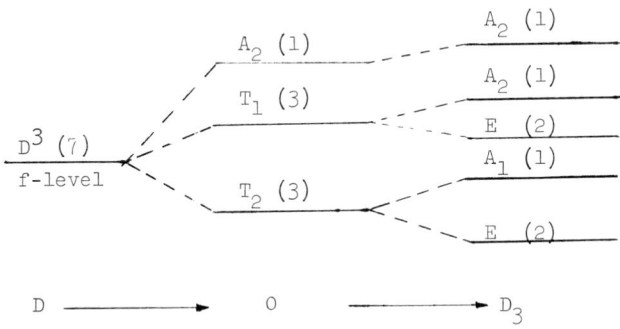

Now let us remove this strain, return to cubic symmetry, and apply a new strain in the (110) direction. This will preserve the two-fold axis in the (110) direction and the one at right angles to it in the ($\bar{1}10$) direction. Furthermore, the 4-fold axis at right angles to the strain direction — that is, the one along (001) — becomes a two-fold axis. Hence the new symmetry elements are three mutually orthogonal two-fold axes, and the group is the D_2 (or 222) group. The character table on page 95 is drawn to show which elements of 0 are also in D_2. (Note that there is no such simple relationship between D_3 and D_2. All elements of D_2 are <u>not</u> in D_3 but they <u>are</u> in 0.) For the sake of labeling, the z-axis is taken along the 4-fold axis in 0 which survives as a two-fold axis in D_2.

This time, the functions of O which transformed there according to A_2 will generate a representation of D_2 whose characters are (1, 1, -1, -1). Similarly T_1 of O→(3, -1, -1, -1) and T_2 of O→(3, -1, 1, 1). The first of these new representations is irreducible, and the others are reducible into the representations $B_1 + B_2 + B_3$ and $A_1 + B_2 + B_3$ of D_2, respectively.

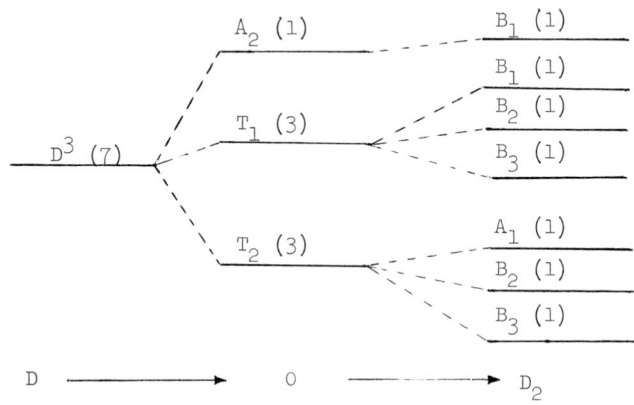

Two things are worth noting here: (1) A loss of symmetry does not <u>require</u> the splitting of degenerate levels; it merely tells the maximum possible splitting that the new symmetry will <u>permit</u>. If this maximum is not achieved under the particular potential, we have accidental degeneracy. (2) The <u>order</u> of levels and <u>amount</u> of splitting is determined by the type and strength of the potential, not by the symmetry.

<u>Particles with Spin</u>. Let us now incorporate the spin of our particles into this formalism. We consider atomic states formed from two electrons so that the multiplet states will have integral spin. The spinors then transform in the same way that spherical harmonics do. Suppose for example that our atom has two valence electrons and we consider the atomic state which in the absence of electrostatic interactions between the electrons would be labeled

2p3d. Now we turn on the interactions. We may represent the wave functions of the new states by linear combinations of products of the type $\Psi_{2p}^{(m)} \Psi_{3d}^{(m')}$. Now $\{\Psi_{2p}^{(m)}\}$ transforms according to the D^1 representation of the full group, and $\{\Psi_{3d}^{(m')}\}$ transforms according to D^2. Hence the product wave functions must transform according to the direct product representation $D^1 \times D^2$. This, as we have seen, may then be reduced into irreducible components. We have also seen on page 75 how to use the character tables to do this. In fact we prove here the following important theorem on the irreducible components of direct products like $D^1 \times D^2$.

  <u>Theorem</u>: Given two irreducible representations of the full rotation group D^{ℓ_1} and D^{ℓ_2}, the irreducible components of the direct product representation $D^{\ell_1} \times D^{\ell_2}$ are $D^{\ell_1+\ell_2}$, $D^{\ell_1+\ell_2-1}$ . . . $D^{|\ell_1-\ell_2|}$. In other words,

$$D^{\ell_1} \times D^{\ell_2} = \sum_{\ell=|\ell_1-\ell_2|}^{\ell_1+\ell_2} D^{\ell} \; ;$$

The sum here is only symbolic. We do not mean the matrix sum but what may be called a "direct sum."

  <u>Proof</u>: We have seen earlier that the characters $\chi^{\ell}(\alpha)$ of the direct product representation of a group are equal to the product of the characters $\chi^{\ell_1}(\alpha)$, $\chi^{\ell_2}(\alpha)$.

$$
\begin{aligned}
\chi^{\ell}(\alpha) &= \chi^{\ell_1}(\alpha) \; \chi^{\ell_2}(\alpha) &\text{(see page 94)}\\[4pt]
&= \sum_{m=-\ell_1}^{\ell_1} e^{im\alpha} \sum_{m=-\ell_2}^{\ell_2} e^{im'\alpha}\\[4pt]
&= \sum_{m=-\ell_1}^{\ell_1} \sum_{m'=-\ell_2}^{\ell_2} e^{i(m+m')\alpha}\\[4pt]
&= \sum_{\ell=|\ell_1-\ell_2|}^{\ell_1+\ell_2} \sum_{M=-\ell'}^{\ell'} e^{iM\alpha}\\[4pt]
&= \sum_{\ell=|\ell_1-\ell_2|}^{\ell_1+\ell_2} \chi^{\ell'}(\alpha)
\end{aligned}
$$

Now since the reduction of a representation is unique, the theorem is proved.

This theorem may be considered to be the proof of the vector model for the addition of angular momenta. When we apply it to our particular example we get

$$D^1 \times D^2 = D^1 + D^2 + D^3 \ .$$

That is, the product wave functions $\{\Psi_{2p}^m \ \Psi_{3d}^{m'}\}$ may be taken in linear combinations which transform under the full rotation group according to one of the $\ell = 1, 2, 3$ irreducible representations of the group. We shall see later that this is equivalent to saying that these linear combinations are eigenfunctions of the angular momentum operator with eigenvalue ℓ. Hence the designation of the new states as P, D, and F states.

The spin of a particle adds a new coordinate dimension to the single particle wave function, and for spin-$\frac{1}{2}$ particles, this coordinate may take but two values, and therefore the wave function may be expressed in the form $\Psi(x,y,z,\sigma_z) \equiv \Psi_+(x,y,z) \ u_+(\sigma_z) + \Psi_-(x,y,z) \ u_-(\sigma_z)$, where $u_+(+1) = 1$, $u_+(-1) = 0$, $u_-(+1) = 0$, $u_-(-1) = 1$. Now since any function of σ_z can be put in this form, then in particular $\hat{P}_R u_+$ and $\hat{P}_R u_-$ must also be expressible as linear combinations of u_+ and u_-. This proves that u_+ and u_- generate a two-dimensional representation of the full rotation group. Later we shall see that it is an irreducible representation called $D^{\frac{1}{2}}$, for which $\chi^{\frac{1}{2}}(\alpha) = \dfrac{\sin\left(\frac{1}{2}+\frac{1}{2}\right)\alpha}{\sin \alpha/2}$.

Now since the spin variable is in a different space from the other coordinates, we must inquire separately how the product functions $u^{\frac{1}{2}} u^{\frac{1}{2}}$ transform. But the theorem on page 99 evidently applies here because the characters are defined exactly the same way. Thus, $D^{\frac{1}{2}} \times D^{\frac{1}{2}} = D^0 + D^1$. And so we have a vector model for spins.

Finally we get, from these group-theoretical arguments, the designations ^1P, ^1D, ^1F, ^3P, ^3D, ^3F from our 2p and 3d electrons. Now when we turn on electrostatic and spin interactions, we expect the central field level,

2p3d, to split into these six levels.

Now let us consider what happens when we add a spin-orbit term of the form $\hat{\underline{\sigma}} \cdot \hat{\underline{L}}$ to the Hamiltonian. The form of the term guarantees that it is invariant under the full rotation group if both spin and space coordinates are transformed together but not otherwise. Since this term removes some symmetry of the Hamiltonian without adding any new symmetries, we expect each of the six levels to split further, and this is exactly what happens.

Each of the degenerate wave functions of the ^3D level for example, is a product of a spin function transforming according to D^1 and a spatial function transforming according to D^2. Now if the full group of rotations is performed simultaneously on both spin and space coordinates, the product wave function must transform according to $D^1 \times D^2 = D^1 + D^2 + D^3$. Thus the ^3D level will split into three levels, each transforming as though it had angular momentum 1, 2, and 3, respectively. These numbers are called the "total angular momentum," j.

Thus by group-theoretical means we have proved that the vector model applies to spin functions and space functions separately and together.

Now let us return to the problem of finding the splittings in the energy levels of our atom when we put it in a crystal field. There are two extreme cases.

Case (1). <u>Spin-orbit effects $\gg$ crystal field effects</u>. For a concrete example we consider a ^3P level in the atom and a crystal field of cubic symmetry. The spin-orbit effect, we have seen, divides this level into $^3P_0, {}^3P_1$, and $^3P_2$ levels. Now each of these levels transforms according to $D^j, j = 0, 1, 2$; so our procedure from this point on is exactly the same as that discussed on pages 94-97. The $^3P_2$ level, for example, transforms as D^2; therefore its eigenfunctions generate a <u>representation of O</u>, whose characters are $(5, 1, 1, -1, -1)$ and which therefore must be reducible into E and T_3. Now simply because $^3P_2$ transforms like D^2 it must be 5-fold degenerate, since $\chi^2(E) = \chi^2(O) = 5$. We find that in the cubic field it splits into a two-fold and a three-fold level.

Case (2). Crystal field effects >> spin-orbit effects. In this case we suppose that the crystal field acts only on the space coordinates of the ^3P eigenfunctions. Thus only spatial symmetry is removed in the crystal field, and we carry along the symbolic 3 to indicate that the spin part transforms according to D^1. The space coordinates transform according to D^1 since L = 1 for a P function. Consequently the ^3P functions generate a representation of O whose characters are (3, -1, -1, 0, 1) and therefore transform as T_1 under O. We label this level $^3T_1$ to remind us of the spin transformation properties.

Now we turn on the spin-orbit potential. This time since the unperturbed eigenfunctions are products of space functions transforming according to T_1 and spin functions transforming according to D^1, the product functions will transform under the operations of O, applied both to spin and space coordinates, according to the direct product of T_1 with the <u>irreducible representations contained within</u> <u>the representation generated by these spin functions under</u> <u>the operations of O.</u> In other words, we seek irreducible representations of O generated by spin and space functions separately, and we form all possible direct product representations. Finally we decompose these into irreducible components to get the new splitting.

Now, since the spin functions transform as D^1, and since, as we have already seen, they therefore transform as T_1 under O, it follows that the $^3T_1$ levels generate, under operations of O acting simultaneously in both spaces, the representation $T_1 \times T_1 = A_1 + E + T_1 + T_2$.

By following this procedure for the ^3P level, we can confirm this diagram. (Degeneracies shown in parenthesis.)

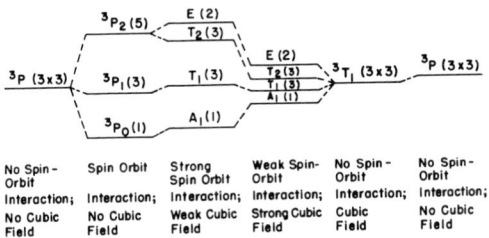

No Spin-Orbit Interaction; No Cubic Field | Spin Orbit Interaction; No Cubic Field | Strong Spin Orbit Interaction; Weak Cubic Field | Weak Spin-Orbit Interaction; Strong Cubic Field | No Spin-Orbit Interaction; Cubic Field | No Spin-Orbit Interaction; No Cubic Field

THE DOUBLE GROUP

When we introduced particles with spin into the problem of crystal field splittings, we restricted the case to atomic states originating from product states of two electrons because such states transformed accordingo to D^1 and that representation was well known to us. Now we consider the states which may arise from an _odd_ number of electrons. As we shall see later, the 2j+1 unit spin functions span a (2j+1)-dimensional subspace which is invariant under D; so they generate, according to theorem (1), page 53 , a representation of it which we shall call D^j. When j is integral, the representations are those generated by $\{Y_\ell^m\}$. But when j is half an odd integer, we obtain new representations with a rather strange property. The matrix which represents a rotation α about some axis is always equal to the _negative_ of the matrix representing a rotation of $\alpha+2\pi$ about the same axis.

There are at least two ways to proceed at this point. We can say that the group D should be thought of as a set of operations performed upon a physical body and that, as such, a rotation by 2π is no rotation at all. Nevertheless, we are faced with the fact that rotations of spinors by α and $\alpha+2\pi$ are represented by two different matrices if $2j$ is odd. From this point of view, we speak of these as "double-valued representations" and generalize the idea of a representation to permit systematic handling of the ambiguity of the sign associated with a matrix representing a physical rotation.

The other point of view maintains the idea of single-valuedness — one group element, one matrix. As a consequence we must accept that rotations by α and $\alpha+2\pi$ are simply not the same element of the group. In other words, the set of operations which we have called D is not a group at all when it operates on half-odd-integral spinors, because it is not closed. On the other hand, rotations of α and $\alpha+4\pi$ are the same element; so the group is closed if it is enlarged by adding for every element $C(\alpha,\underline{r})$ of D a new element $C(\alpha+2\pi,\underline{r})$. We shall label the new group D' and call it the "double group of full rotations," a nomenclature which reflects this second point of view. The advantage of the second approach is that the new group D' has single-valued representations; so the representation theory which we have developed may be applied to it without further thought.

Either approach seems very artificial, very un-physical; but this stems not from the group theory but from the very concept of spin. Consider the logical steps: (1) The spectral lines of sodium are doublets. (2) This suggests an extra degree of freedom in the atomic states and an interaction. (3) The spectra can be understood by means of a term diagram whose general structure may be derived by supposing that the extra degree of freedom is quantized, adds to itself according to the vector model for addition of angular momentum, adds to orbital angular momentum in the same way, leads to the same selection rules as those for orbital angular momentum, and is subject to an interaction with the orbital angular momentum. (4) If it is an intrinsic angular momentum of the electron, then its

eigenfunctions must generate a representation of the full
rotation group. (5) The fact of splitting into doublets
means that the representation is two-dimensional. (6)
The two-dimensional representation is $D^{\frac{1}{2}}$ and is double-
valued.

To avoid these apparent artificialities we would require rather revolutionary changes in the theory of spin.

Now let us turn to the task of constructing character
tables for the double group of full rotations and also for
the double groups generated in the same way from any of the
finite rotation groups we have discussed. Let G be the
group of actual rotations R and G' be the double group.
First we prove that G and G' are homomorphic. Let $\hat{P}_R$, $\hat{P}_S$,
and $\hat{P}_T$ be operations of G for which RS = T. Now if $\underline{u}$ is a
set of __spinors__ spanning an invariant subspace under G, then
we can write $\hat{P}_R \underline{u} = \pm \underline{u} \, \Gamma(R)$, and similarly for $\hat{P}_S$ and $\hat{P}_T$.
On the other hand, $\hat{P}_T \underline{u} = (\hat{P}_R \hat{P}_S) \underline{u} = \hat{P}_R (\hat{P}_S \underline{u}) =$

$\hat{P}_R [\pm \underline{u}\Gamma(S)] = \hat{P}_R \underline{u} [\pm \Gamma(S)] = [\pm \underline{u}\Gamma(R)] [\pm\Gamma(S)]$. Hence

$[\pm\Gamma(R)] [\pm\Gamma(S)] = \pm\Gamma(T)$ if RS = T. But the matrices

$\{\pm\Gamma(R)\}$ for all R in G may be taken themselves as a group
element of G'. If now we associate with both matrices
$+\Gamma(R)$ and $-\Gamma(R)$ of G' the rotation R of G, the homomorphism
follows at once.

In consequence of the homomorphism between G and G'
we conclude that the irreducible representations of G are
also irreducible representations of G', although, of course,
they can never be faithful representations of G'.

Now we inquire into the class structure of G'. Homomorphism alone requires that one or more classes of G'
shall map into one class of G. Now suppose that S and T
belong to the same class of G. This means that for some R in
G: $RSR^{-1} = T$. But this implies that $[\pm\Gamma(R)] [\pm\Gamma(S)]$
$[\pm\Gamma(R^{-1})] = \pm \Gamma(T)$. It is easy to show that $\Gamma(R^{-1}) =$
$\pm \Gamma^{-1}(R)$, and since the sign ambiguities are independent of
one another, we can write $\Gamma(R)\Gamma(S)\Gamma^{-1}(R) = \pm \Gamma(T)$. That is,

$\Gamma(S)$ is in the class either of $\Gamma(T)$ or $-\Gamma(T)$, and $-\Gamma(S)$ is in the class of the other one. There is, therefore, the possibility of <u>two</u> classes of G' mapping into one class of G. On the other hand, some $\Gamma(R)$ may exist which puts $\Gamma(S)$ into the same class with $-\Gamma(S)$; then we would have only <u>one</u> class of G' mapping into one class of G. A more careful analysis than ours shows that the latter case arises if, and only if, S is a two-fold rotation in G and there is another two-fold axis perpendicular to the axis of S.

Thus far we have spoken of the matrices $\pm\Gamma(R)$ as the elements of G', but we may as well speak of some operations as forming the elements. Thus we let R' be the rotation represented by one of the matrices $\pm\Gamma(R)$ and $\overline{R}'$ be the other. These operations will be rotations which differ by 2π. The two operations E' and $\overline{E}'$ are those represented by $\pm\Gamma(E)$. We choose E' to be that one of the pair which is represented in the irreducible representations by the unit matrix I. Then $\overline{E}$ must be represented by $-I$ in the <u>faithful</u> representations. E' is a rotation by $4n\pi$ and $\overline{E}'$ is a rotation by $(4n+2)\pi$. E' and $\overline{E}'$ commute with all other operations and $\overline{E}^2 = E$. For the doubling of the rest of the elements of G, there is still some arbitrary labeling. For example, consider the abelian group C_3 whose elements are generated by repeated application of the rotation C_3. We can generate the double group C_3' by repeated application of an operation we call C_3'. But whereas $(C_3)^3 = E$ in C_3, $(C_3')^3 = \overline{E}'$ in C_3'. Let us list the elements of C_3' in the following order: $\{E'=(C_3')^6, C_3', (C_3')^2, \overline{E}'=(C_3')^3, (C_3')^4, (C_3')^5\}$; and let us list the elements of C_3 in the order $\{E = (C_3)^3, C_3, C_3^2 =(C_3)^2\}$. Now if we think of C_3 and C_3' as both representing rotations of $\frac{2\pi}{3}$, then we are led to the following labeling of the elements of C_3', in the same order as before: $\{E', C_3', C_3^{2'}, \overline{E}', \overline{C}_3' = \overline{E}'C_3', \overline{C}_3^{2'} = \overline{E}'C_3^{2'}\}$. On the other hand, if we think of C_3^2 not as a rotation by $\frac{4\pi}{3}$ but by $\frac{-2\pi}{3}$, then we are more apt to label it C_3^{-1}, and this leads us to a similar kind of labeling in the double group $\{E', C_3', \overline{C}_3^{-1} = \overline{E}'C_3^{-1'}, \overline{E}', \overline{C}_3' = \overline{E}'C_3', C_3^{-1'}\}$.

In one case, $(C_3')^2$ is labeled by an unbarred symbol and $(C_3')^5$ by a barred symbol. In the other, the barring is transposed. Either is self-consistent. Which is preferred? For the abelian group discussed it makes no difference, but suppose the three-fold axis is in the D_3 group. Then C_3 and C_3^2 fall into the same class. In the double group D_3', the four elements C_3', $(C_3')^2$, $(C_3')^4$, and $(C_3')^5$ will, as we have seen, go into two classes. Then it is convenient to label so that all barred elements fall into one class and all unbarred elements fall into the other. This is always possible since R' and $\overline{E}$ R' = $\overline{R}'$ always, with the one exception on page 106, fall into separate classes.

Let us sum up now the following rules for double group representations. Let G be the group of rotations R, and let G' be the double group with elements R' and $\overline{R}'$.

(1) Every irreducible representation of G is also an irreducible representation of G'. These are called "single-valued representations of G."

(2) If elements {R} form a class in G, then the corresponding elements {R'} form a class in G' and the corresponding elements {$\overline{R}'$} form a distinct class in G' unless rule 3 applies.

(3) If R is a two-fold rotation in G,and G also contains another two-fold axis perpendicular to the axis of R, then the elements {R', $\overline{R}'$} form a single class of G'.

(4) The difference between the number of classes of G' and of G is the number of new irreducible representations of G'.

(5) In the new irreducible representations, $\chi^{(i)}(\overline{E}') = -\chi^{(i)}(E') = -\ell_1$. And in general $\chi^{(i)}(\overline{R}') = -\chi^{(i)}(R')$. This is true because $\Gamma^{(i)}(\overline{E}') = -\Gamma^{(i)}(E') = -I$ and $\overline{R}' = \overline{E}'R'$. For this reason these are called the "double-valued representations of G," although they are, of course, single-valued representations of G'.

(6) In the exceptional case of rule 3 we have, in addition to rule 5, the fact that $\chi(\overline{C}_2') = \chi(C_2')$. Hence $\chi(C_2') = 0$.

(7) If G has an n-fold axis, with elementary operation C_n, then a one-dimensional double-valued representation of C_n must equal an n$\underline{\text{th}}$ roots of -1. This is true since $[\Gamma^{(1)}(C_n')]^n = \Gamma(\overline{E}') = -1$ for one-dimensional new representations.

(8) Hence in the exceptional case of rule 3 there can be no one-dimensional representations.

Examples:

(A) Let G=D_2 (or 222) and G'=D_2'. h=4, so h'=8. By rule 3 we get only one new class, that of $\overline{E}$. (We now drop the prime in labeling elements of the double groups.) By rule 4 we get one extra irreducible representation. Rule 8 tells us it must be at least two-dimensional. By rule 1 we get the four single-valued irreducible representations by copying the four irreducible representations we found on page 59 for D_2. Now in D_2' $\sum_{i=1}^{5}\ell_i^2=8$; therefore the new representation must in fact be two-dimensional. By rule 5 we get $\chi(\overline{E})=-2$ in the new representation. And by rule 6 we get zero elsewhere in the new representation. Thus the character table of D_2' is established

	E	$\overline{E}$	C_{2z} , $\overline{C}_{2z}$	C_{2y} , $\overline{C}_{2y}$	C_{2x} , $\overline{C}_{2x}$
A_1	1	1	1	1	1
B_1	1	1	1	-1	-1
B_2	1	1	-1	1	-1
B_3	1	1	-1	-1	1
E'	2	-2	0	0	0

(B) Let $G = D_3$ (or 32) and $G' = D_3'$. $h = 6$, so
$h' = 12$. According to rule 2, we get a new class for every
class of D_3 — a total of three new classes. Hence, by
rule 4 we must have three new irreducible representations.
By rule 1 we get the three single-valued irreducible representations by copying the three irreducible representations
we found for D_3 on page 64 . Now $\sum_{i=1}^{6} \ell_i^2 = 12$; but
$\sum_{i=1}^{3} \ell_i^2 = 6$, so $\sum_{i=4}^{6} \ell_i^2 = 6$. This only has proper solutions
if the new representations are of dimension 1, 1, and 2.
This gives $\chi^{(i)}(E)$; rule 5 gives $\chi^{(i)}(\overline{E})$, $i = 4, 5, 6$.
Rule 7 tells us that for the one-dimensional representa-
tions $\chi^{(i)}(C_2) = \pm i$. Rule 5 says $\chi^{(i)}(\overline{C}_2)$ must have the
opposite sign. Similarly $\chi^{(i)}(C_3) = -\chi^{(i)}(\overline{C}_3) = -1, -\omega,$
$-\omega^2$, where $\omega = -\frac{1}{2} + \frac{\sqrt{3}}{2} i$. To complete the character table,
we may use the orthogonality properties. The row ortho-
gonality relation, page 33 , for G' is

$$\sum_k \chi^{(i)}(C_k)^* \, \chi^{(j)}(C_k) \, N_k = h' \, \delta_{ij} \ .$$

Now we may sum over barred classes and unbarred classes
separately. Since $\chi^{(i)}(\overline{C}_k)^* \, \chi^{(j)}(\overline{C}_k) = [-\chi^{(i)}(C_k)^*] \times$
$[-\chi^{(j)}(C_k)]$ and N_k is the same for both types, we may take
the sum over either type alone and it must equal $\frac{h'}{2} \delta_{ij}$.
But $\frac{h'}{2} = h$. We have assumed here that i and j both
labeled the new, double-valued representations. If both
are the old single-valued types, the conclusion also
follows. If one is new and one is old, then the orthogonality is assured independent of the orthogonality relation. We put these results in rule 9 below.

Now consider the column orthogonality relation, page
34.   There are three distinct cases which arise.

$$\sum_i \chi^{(i)}(C_k)^* \, \chi^{(i)}(C_\ell) = \frac{h'}{N_k} \delta_{k\ell}$$

We may break this into a sum over the single-valued and one
over the double-valued representations. If $C_\ell = C_k$ or if

$C_\ell = \overline{C}_k$, then the sum over single-valued representations is equal to $\frac{h}{N_k}$. Since h' = 2h, the sum over the double-valued representations must give $\frac{h}{N_k}$ if $C_\ell = C_k$ and $(-\frac{h}{N_k})$ if $C_\ell = \overline{C}_k$. If $\overline{C}_k \neq C_\ell \neq \overline{C}_k$, then the whole sum vanishes —— as it did when $\overline{C}_k = C_\ell$-but so does the sum over the single-valued representations. So the sum over the new ones must also vanish. Thus we get the new rules, with the convention that C_k is a class of unbarred elements and $\overline{C}_k$ is the corresponding class of barred elements of G'. Recall that h is the order of G, not G'.

$$(9) \quad \sum_k \chi^{(i)}(C_k)^* \; \chi^{(j)}(C_k) \; N_k =$$

$$\sum_k \chi^{(i)}(\overline{C}_k)^* \; \chi^{(j)}(\overline{C}_k) \; N_k = h \; \delta_{ij}$$

for i and j labeling new representations.

$$(10) \quad \sum_i \chi^{(i)}(C_k)^* \; \chi^{(i)}(C_\ell) = \frac{h}{N_k} \; \delta_{k\ell}$$

if the sum is over the new representations only.

$$(11) \quad \sum_i \chi^{(i)}(C_k)^* \; \chi^{(i)}(\overline{C}_\ell) = -\frac{h}{N_k} \; \delta_{k\ell}$$

if the sum is over new representations only.

Now let us return to the construction of the character table of D_3'. We saw that $\chi^{(j)}(C_2) = \pm i$ for the new one-dimensional representations. Since h = 6 and there are 3 C_2 operations in D_2' (or D_2, for that matter), it follows from rule 10, with $C_k = C_\ell =$ class of C_2, that $\chi(C_2) = 0$ in the two-dimensional, double-valued representation. Similarly by rule 10, with $C_k =$ class of E and $C_\ell =$ class of C_2, we find that $\chi(C_2) = i$ in one one-dimensional new representation, and $\chi(C_2) = -i$ in the other. This completes the C_2 and $\overline{C}_2$ column of the table. By a similar set of arguments we can find the characters of C_3 and $\overline{C}_3$ in the new representations. The table is given below.

	E	$\overline{E}$	$3C_2$	$3\overline{C}_2$	$2C_3$	$2\overline{C}_3$
A_1	1	1	1	1	1	1
A_2	1	1	-1	-1	1	1
E	2	2	0	0	-1	-1
A_1'	1	-1	i	$-i$	-1	1
A_2'	1	-1	$-i$	i	-1	1
E'	2	-2	0	0	1	-1

These examples suggest that the construction of the character table for the double group is not more difficult, and is frequently simpler, than that of the original. This is so in fact.

Before we return to the application of this formalism to the problem of crystal field splittings, one remark is in order. Since the double group G' has a set of elements {R} and a corresponding set {$\overline{R} \equiv R\overline{E}$} and since E and $\overline{E}$ form a group, and since $\overline{E}$ commutes with all R, isn't the double group really a product group? The answer is "No" because the set of elements {R} is not a group. Thus $(C_n)^n = \overline{E}$, and $\overline{E}$ belongs to the set {$\overline{R}$} and not to the set {R}.

Let us now return to the problem of crystal field effects on the energy levels of an atom. The procedure is exactly the same as that used on pages 94-103 except that we must have now the double-valued representations of every group we encountered before. (The character table of D', the double full rotation group is generated from the formulae, $\chi^J(\alpha) = \frac{\sin(j+\frac{1}{2})\alpha}{\sin\alpha/2}$, $\chi^J(\overline{\alpha}) = \chi^J(\alpha+2\pi)$.) Suppose, for example, that residual electrostatic and spin-spin interactions produce a ^4F level in a three-electron atom. Now we introduce spin-orbit forces. The new wave functions are to be expressed as products of a spinor which transforms under D' according to $D^{3/2}$ and a coordinate function which transforms according to D^3. Hence we form $D^{3/2} \times D^3$ and then reduce the result. We have seen that this gives the states which a vector model would predict. One of these states, a $^4F_{9/2}$ transforming under D' according to $D^{9/2}$, may then be studied as an example of the effects of the crystal field.

For a full study we would need a table of the kind given on page 95 , but now for the double groups. We shall not work this out in detail here but merely sketch in the logical steps and remark on a few general results.

When we turn on the crystal field we remove some of the symmetry of D' and go, let us say, to O'. The ten degenerate eigenfunctions of the $^4F_{9/2}$ state will generate a representation of O' which in general is reducible. From the character table of D' we may find the characters of this representation of O' and then may reduce it as in the previous example on page 96 . Then, by stressing, we may go to lower and lower symmetries.

But suppose the crystal field effects were much stronger than the spin-orbit effects. Then we must consider the crystal field splittings first. The ^4F level is then thought of as a highly degenerate set of states which transform according to D^3. Therefore when we turn on the crystal field these functions will generate a representation of O' which may be determined by an inspection of the characters of the surviving symmetry elements in the representation D^3. (We follow here exactly the same arguments as given in case (2), page 102 .) This representation of O' may be reduced, and one of its components is T_2. Thus one of the states split off from ^4F is labeled $^4T_2$, which indicates the spin degeneracy.

Now when we turn on the spin-orbit forces, the spin functions come under the influence of the crystal symmetry. So first we must find the representations of O' which they generate. The fact that F is a quartet state means that the spinor part transforms according to $D^{3/2}$; so we look at the characters of those elements of D' which survive in O' and which belong to $D^{3/2}$ of D'. This is the character of a representation of O' which happens to be Γ_8 in one of the notations. Now to find the spin-orbit levels we must take the direct product of Γ_8 with T_2 (also called Γ_5 in that notation) and reduce the resulting representation. The reason is exactly that given on page 102.

Three important general results of this procedure should be emphasized. (1) When we go from higher to lower symmetry, the degenerate wave functions which transformed according to a single- (or double-) valued representation, will generate purely single- (or double-) valued representations of the group of lower symmetry, and these may be reduced into purely single- (or double-) valued irreducible components. (2) Two sets of eigenfunctions may generate a direct product representation of a group; if both sets transform according to a single-valued, or both according to a double-valued, representation, then the irreducible components of the direct product representation will be purely single-valued — otherwise, purely double-valued. (3) The direct product representation $\Gamma_i \times \Gamma_j$ contains the identical representation Γ_1 if, and only if, $i = j$.

Before we leave the problem of spin splittings we should mention the effect of considering groups with inversion symmetry or other improper rotations. Inversion symmetry is easily incorporated into our scheme by the device of the direct product of groups discussed on page 71. The double group is formed first, then the direct product is taken with $\{E, i\}$. On the other hand, the inversion element might as well be left out in our problem, since stressing never changes it. (When we come to the calculation of transition rates, however, we shall find it very important.)

Other improper rotation groups are isomorphic with proper rotation groups; so no difficulty in forming double groups should be found there either.

PERTURBATION THEORY AND TRANSITION PROBABILITIES

Perturbation theory and transition probabilities are two problems which are immensely simplified (1) if we take full advantage of any symmetries which may be present and (2) if the problem of symmetry is handled within the framework of group theory. In both of these problems, and in many others, the greater part of the labor goes into the calculation of matrix elements. Any light which can be thrown on this subject by the theory of symmetry will be most helpful.

Recall that in theorem (vii), page 54, we proved a result which, in the language of quantum mechanics, stated that wave functions which transform according to two different irreducible representations of some symmetry group of unitary operators were necessarily orthogonal. It is perhaps worth reviewing that theorem now, since our purpose will be to generalize it and to treat not merely $(\Psi_\varkappa^{(1)}, \Phi_\lambda^{(j)})$, but also $(\Psi, \hat{O}\Phi)$, where $\hat{O}$ is some operator.

Since, by theorem (viii), page 55, Ψ and Φ can be expressed as linear combinations of basis functions spanning irreducible subspaces invariant under the group G of symmetry operations appropriate to our problem, we have lost no generality if we prove theorems involving only those basis functions. That is, instead of treating $(\Psi, \hat{O}\Phi)$, we need consider only $(\Psi_\varkappa^{(1)}, \hat{O}\Phi_\lambda^{(j)})$ in the new theorems. Note that by calling one funtion Ψ and the other one Φ, we mean that $\Psi_\varkappa^{(1)}$ is not necessarily equal to $\Phi_\varkappa^{(1)}$. Thus, for example, $\Psi_\varkappa^{(1)}$ might be a linear basis function and $\Phi_\varkappa^{(1)}$ a quadratic basis function, both transforming according to the $i^{\underline{th}}$ irreducible representation of G. Or they might be different basis functions in the same invariant subspace: $\underline{\Psi}^{(1)} = (x,y)$, $\underline{\Phi}^{(1)} = (x+y, x-y)$. The point is that though they both transform according to the $i^{\underline{th}}$ irreducible representation, the matrices which they generate are not necessarily identical, although they must be equivalent, of course.

Notation: Let the sets $\{\Psi_{\varkappa}^{(i)}\} \equiv \underline{\Psi}^{(i)}$ and $\{\Phi_{\lambda}^{(j)}\} = \underline{\Phi}^{(j)}$ be ortho-normal basis vectors of the $i\underline{th}$ and $j\underline{th}$ irreducible subspaces, respectively, invariant under G. Let the irreducible representation which they therefore generate be defined by the identities,

$$\hat{P}_R\underline{\Psi}^{(i)} \equiv \underline{\Psi}^{(i)} \, \Gamma_{\Psi}^{(i)}(R) \quad \text{and} \quad \hat{P}_R \, \underline{\Phi}^{(j)} \equiv \underline{\Phi}^{(j)} \, \Gamma_{\Phi}^{(j)}(R) \; .$$

Let $\hat{O}$ be any operator for which $(\Psi, \hat{O}\Phi)$ is defined.

With this notation we add some more vector algebra theorems to those proved on pages 53-55.

Theorem ix: If all unitary operations $\{\hat{P}_R\}$ of G commute with $\hat{O}$ — that is, they leave $\hat{O}$ invariant — then $(\Psi_{\varkappa}^{(i)}, \hat{O}\Phi_{\lambda}^{(j)}) = 0$ if $i \neq j$. If $i = j$ and if $\Gamma_{\Psi}^{(i)} = \Gamma_{\Phi}^{(i)}$, then $(\Psi_{\varkappa}^{(i)}, \hat{O}\Phi_{\lambda}^{(i)}) = \sum_{\mu} (\Psi_{\mu}^{(i)}, \hat{O}\Phi_{\mu}^{(i)}) \dfrac{\delta_{\varkappa\lambda}}{\ell_i}$.

Proof: The proof follows the lines of theorem (vii), page 54 , and we use the same vector and matrix notation as there.

$$(\underline{\widetilde{\Psi}}^{(i)}, \hat{O}\underline{\Phi}^{(j)}) = (\hat{P}_R\underline{\widetilde{\Psi}}^{(i)}, \hat{P}_R\hat{O}\underline{\Phi}^{(j)}) \quad \text{(by unitarity of } \hat{P}_R\text{)}$$

$$= (\hat{P}_R\underline{\widetilde{\Psi}}^{(i)}, \hat{O}\hat{P}_R\underline{\Phi}^{(j)}) \quad \text{(by commutativity with O)}$$

$$= (\underline{\widetilde{\Psi}^{(i)}\Gamma_{\Psi}^{(i)}}(R), \hat{O}\underline{\Phi}^{(j)}\Gamma_{\Phi}^{(j)}(R)) \quad \text{(by definition)}$$

$$= (\widetilde{\Gamma_{\Psi}^{(i)}(R)}\underline{\widetilde{\Psi}}^{(i)}, \hat{O}\underline{\Phi}^{(j)}\Gamma_{\Phi}^{(j)}(R)) \quad \text{(transpose of products)}$$

$$= \widetilde{\Gamma_{\Psi}^{(i)*}}(R)(\underline{\widetilde{\Psi}}^{(i)}, \hat{O}\underline{\Phi}^{(j)})\Gamma_{\Phi}^{(j)}(R) \quad \text{(definition of inner product)}$$

$$\longrightarrow \quad = \Gamma_{\Psi}^{(i)}(R)^{\dagger}(\underline{\widetilde{\Psi}}^{(i)}, \hat{O}\underline{\Phi}^{(j)})\Gamma_{\Phi}^{(j)}(R) \quad \text{(definition of hermitian matrix)}$$

$$= (\Gamma_{\Psi}^{(i)}(R))^{-1}(\underline{\widetilde{\Psi}}^{(i)}, \hat{O}\underline{\Phi}^{(j)})\Gamma_{\Phi}^{(j)}(R)) \quad \text{(theorem (iv) page 54)}$$

$$\Gamma_{\Psi}^{(i)}(R)(\underline{\Psi}^{(i)}, \hat{O}\underline{\Phi}^{(j)}) = (\underline{\Psi}^{(i)}, \hat{O}\underline{\Phi}^{(j)}) \, \Gamma_{\Phi}^{(j)}(R) \quad \text{(left multiplication)}$$

Thus. $(\underline{\Psi}^{(i)}, \hat{0}\underline{\Phi}^{(j)})$ is a matrix which satisfies the conditions of Theorem III, page 26 . Now if $i \neq j$, then $\Gamma_{\Psi}^{(i)}$ is certainly not equivalent to $\Gamma_{\Phi}^{(j)}$. Hence, by the corollary on page 28 either $\ell_i \neq \ell_j$ or else $(\underline{\Psi}^{(i)}, \hat{0}\underline{\Phi}^{(j)})$ is a null matrix. But if $\ell_i \neq \ell_j$, then Theorem III itself requires it to be a null matrix. This proves the case for $i \neq j$.

If now $i = j$ and also $\Gamma_{\Psi}^{(i)}(R) = \Gamma_{\Phi}^{(i)}(R) \equiv \Gamma^{(i)}(R)$, we may prove the second case most easily by writing out the components of the equation indicated above by an arrow.

$$(\Psi_{\varkappa}^{(i)}, \hat{0}\Phi_{\lambda}^{(i)}) = \sum_{\mu,\nu=1}^{\ell_i} \Gamma^{(i)}(R)_{\mu\varkappa} (\Psi_{\mu}^{(i)}, \hat{0}\Phi_{\nu}^{(i)}) \Gamma^{(i)}(R)_{\nu\lambda}$$

We sum both sides over all R in G and use the Orthogonality Theorem, page 29 , to get

$$(\Psi_{\varkappa}^{(i)}, \hat{0}\Phi_{\lambda}^{(i)}) \, h = \sum_{\mu,\nu=1}^{\ell_i} (\Psi_{\mu}^{(i)}, \hat{0}\Phi_{\nu}^{(i)}) \frac{h}{\ell_i} \delta_{\mu\nu} \delta_{\lambda\varkappa}$$

$$(\Psi_{\varkappa}^{(i)}, \hat{0}\Phi_{\lambda}^{(i)}) = \sum_{\mu=1}^{\ell_i} (\Psi_{\mu}^{(i)}, \hat{0}\Phi_{\mu}^{(i)}) \frac{\delta_{\lambda\varkappa}}{\ell_i} \, , \text{ Q.E.D.}$$

Discussion: If $\hat{0} = \hat{H}$ is the Hamiltonian of a system and G is the group of the Schrödinger equation, then there are no matrix elements of $\hat{H}$ between wave functions which transform according to inequivalent irreducible representations of G. If the wave functions transform according to equivalent, but distinct, irreducible representations, then nothing can be said. It is wise, therefore, to choose new basis functions in one of the invariant subspaces so that the representations become identical with one another. Then the second part of the theorem will guarantee further vanishing matrix elements.

Now let us consider the case in which $\hat{0}$ is not invariant under G. Then it may be expressed, after the fashion of theorem (viii), page 55 , as a linear combination of operations which each transform according to some irreducible representation of G. Let $\hat{0}_{\mu}^{(k)}$ be such an operation and define $\Gamma_0^{(k)}(R)$ such that

$$\hat{P}_R \hat{0}_{\mu}^{(h)} \equiv \sum_{\nu} \hat{0}_{\nu}^{(k)} \Gamma_0^{(k)}(R)_{\nu\mu} \, .$$

Theorem x: Subject to these conditions,
$(\Psi_\varkappa^{(1)}, \hat{O}_\lambda^{(j)} \Phi_\mu^{(k)}) = 0$ unless the direct product representation $\Gamma^{(j)} \times \Gamma^{(k)}$ contains the irreducible component $\Gamma^{(1)}$.

Proof:

$$(\Psi_\varkappa^{(1)} \hat{O}_\lambda^{(j)} \Phi_\mu^{(k)}) = (\hat{P}_R \Psi_\varkappa^{(1)}, \ \hat{P}_R (\hat{O}_\lambda^{(j)} \Phi_\mu^{(k)})) \text{ by unitarity}$$

$$= \sum_{\nu, \eta, \xi} (\Psi_\nu^{(1)} \Gamma_{\nu\varkappa}^{(1)}(R), \hat{O}_\eta^{(j)} \Phi_\xi^{(k)} \Gamma^{(j)}(R)_{\eta\lambda} \Gamma^{(k)}(R)_{\xi\mu}) \text{ by definition}$$

Thus we see that $\hat{O}_\lambda^{(j)} \Phi_\mu^{(k)}$ are basis functions of a direct product space and that they transform according to a direct product representation of G. $\Gamma_{\eta\xi, \lambda\mu} = \Gamma_{\eta\lambda}^{(j)} \Gamma_{\xi\mu}^{(k)}$, or $\Gamma = \Gamma^{(j)} \times \Gamma^{(k)}$. But, as we have seen on pages 74-75 Γ is generally reducible. That is, $\Gamma = \sum_r a_r \Gamma^{(r)}$, where this means a direct sum. But such a reduction of Γ is achieved by a new choice of basis vectors in the product space. These new vectors will be linear combinations of $\{\hat{O}_\lambda^{(j)} \Phi_\mu^{(k)}\}$ for $\lambda = 1, 2 \cdots \ell_j$, $\mu = 1, 2 \cdots \ell_k$. And these new vectors span a space which has been reduced, in the sense that its first three elements, let us say, will only transform into one another under G, the next five will only transform into one another, and so forth. Thus we may write the new vectors as $\hat{O}^{(j)} \Phi^{(k)} = \sum_s a_s \Phi^{(s)}$, where we again mean the direct sum. Hence $(\Psi_\varkappa^{(1)}, \hat{O}_\lambda^{(j)} \Phi_\mu^{(k)}) = \sum_s a_s (\Psi_\varkappa^{(1)}, \Phi_\lambda^{(s)})$. By theorem (ix) this is zero unless $a_1 \neq 0$. This proves the theorem.

Corollaries: (a) In theorem (ix), we put $\hat{O} = 1$, and we get a slightly more general version of theorem (vii) page 54. Wave functions which transform according to inequivalent irreducible representations of G are orthogonal. Wave functions which transform according to equivalent and identical irreducible representations are orthogonal unless they transform according to the same column of the representation matrices.

(b) Set $\hat{O} = 1$ and $\Psi_\varkappa^{(1)} = \Psi_\varkappa^{(1)} = 1$. Interpret the inner product as the usual integral $(f,g) = \int f^*g \, d^{3N}v$. Theorem (ix) says that $\int \Phi^{(1)} d^{3N}v = 0$ unless $i = 1$, that is, unless $\Phi^{(1)}$ is itself invariant under the operations of G. This result has the immediate consequence that the integral of an arbitrary function is equal to the integral of its component which transforms according to the identical representation of G.

Before we go ahead with examples, let us prove a theorem which is somewhat related to these.

Theorem xi: (Generalized Unsöld's Theorem). If the set of orthogonal vectors $\{\phi_\varkappa^{(i)}\} \equiv \underline{\phi}^{(i)}$ are ortho-normal basis vectors of the ith irreducible subspace invariant under G, then $F \equiv \sum_\varkappa \phi_\varkappa^{(i)*}\phi_\varkappa^{(i)} \equiv \underline{\phi}^{(i)*}\underline{\widetilde{\phi}}^{(i)}$ is itself invariant under G.

Proof:

$\hat{P}_R F = \hat{P}_R(\underline{\phi}^{(i)*}\underline{\widetilde{\phi}}^{(i)})$

$= \underline{\phi}^{(i)*}\Gamma^{(i)}(R)^* \underbrace{\underline{\phi}^{(i)}\Gamma^{(i)}(R)}$ (by definition)

$= \underline{\phi}^{(i)*} \Gamma^{(i)}(R)^* \widetilde{\Gamma^{(i)}(R)} \underline{\widetilde{\phi}}^{(i)}$ (transpose of product)

$= \underline{\phi}^{(i)*} [\Gamma^{(i)}(R)\Gamma^{(i)}(R)^\dagger]^* \underline{\widetilde{\phi}}^{(i)}$ (definition of hermitian conjugate)

$= \underline{\phi}^{(i)*} [\Gamma^{(i)}(R) \Gamma^{(i)}(R)^{-1}]^* \underline{\widetilde{\phi}}^{(i)}$ (unitarity of representation)

$= \underline{\phi}^{(i)*} \underline{\phi}^{(i)}$ $(AA^{-1} = I)$

$= F$ (by definition) Q.E.D.

This is a generalization of Unsöld's Theorem for the spherical harmonics, which says that $\sum_{m=-\ell}^{\ell} |Y_\ell^m(\theta,\varphi)|^2$ is invariant under all rotations. Since the set of all rotations are the symmetry elements of the sphere, we say that the closed shells of the atom are spherically symmetric. The generalized theorem implies that the closed shells of systems of any symmetry have precisely that symmetry themselves. A fully occupied T_1 level in a system of cubic symmetry, for example, would have the full symmetry of the cube. Since, as we have seen in the examples, a reduction of symmetry always takes a closed shell into a set of closed shells - the total degeneracy is the same - we may think of the loss of symmetry as affecting whole shells, each of which has the full symmetry of its respective group.

Before we go to applications of these theorems, let us consider how to formulate theorem (ix) if the group in question is a direct product group. This case arises, for example, in systems with inversion symmetry and so is of great practical interest. Of course we could in any case proceed as though we did not know that the group might be analyzed into a direct product; but a great simplification occurs when we use this knowledge.

First, it is quite evident that theorem (viii), page 55, may be generalized to the case of direct product groups. We first decompose with respect to the operations $\{\hat{P}_R\}$ of G and then we decompose those components with respect to $\{\hat{P}_{R'}\}$ of G'. The decomposition is evidently unique if $\hat{P}_R$ commutes with $\hat{P}_{R'}$ for all R and all R'. We express this by saying that there exists an ortho-normal set of vectors $\{\phi_{\varkappa\lambda}^{(ij)}\}$ such that an arbitrary function ϕ can be written

$$\phi = \sum_{ij} \sum_{\varkappa=1}^{\ell_i} \sum_{\lambda=1}^{\ell_j} \phi_{\varkappa\lambda}^{(ij)} \, C_{\varkappa\lambda}^{(ij)}, \text{ where}$$

$$\hat{P}_{RR'} \, \phi_{\varkappa\lambda}^{(ij)} = \sum_{\mu=1}^{\ell_i} \sum_{\nu=1}^{\ell_j} \phi_{\mu\nu}^{(ij)} \, \Gamma^{(ij)}(RR')_{\mu\nu, \varkappa\lambda}$$

As we have seen on page 72, $\Gamma^{(ij)}(RR')_{\mu\nu, \varkappa\lambda} = \Gamma^{(i)}(R)_{\mu\varkappa} \, \Gamma^{(j)}(R')_{\nu\lambda}$ is a unique irreducible representation of G x G'.

Consequently when it comes to proving matrix element theorems, it is sufficient to consider only elements of the form $(\Psi_{\varkappa\lambda}^{(ij)}, \hat{O} \, \phi_{\mu\nu}^{(k\ell)})$. Now the set $\{\phi_{\varkappa\lambda}^{(ij)}\}$ may be symbolized as a row matrix $\underline{\phi}^{ij}$ whose ℓ_i x ℓ_j elements are the individual vectors $\phi_{\varkappa\lambda}^{(ij)}$. Similarly $\Gamma^{(ij)}(RR')$ is an $\ell_i \times \ell_j$ by $\ell_i \times \ell_j$ square matrix. Hence we may use the same compact notation as before: $\hat{P}_{RR'}\underline{\phi}^{(ij)} = \underline{\phi}^{(ij)}\Gamma^{(ij)}(RR')$.

   Theorem xii: If $\{\Psi_{\varkappa\lambda}^{(ij)}\} \equiv \underline{\Psi}^{ij}$ is a set of ortho-
normal basis vectors of an irreducible subspace invariant
under G x G' which transform according to $\Gamma_{\Psi}^{(i)}(R)$ of G and
$\Gamma_{\Psi}^{(j)}(R')$ of G', and if $\underline{\phi}^{(k\ell)}$ is similarly defined, and if
$\hat{P}_{RR'}$ commutes with $\hat{O}$ for all R and R', the $(\Psi_{\varkappa\lambda}^{(ij)},\hat{O}\phi_{\mu\nu}^{(k\ell)}) = 0$
if $i \neq k$ or if $j \neq \ell$. If $i = k$ and $j = \ell$ and if
$\Gamma_{\Psi}^{(i)}(R) = \Gamma_{\phi}^{(i)}(R)$ and $\Gamma_{\Psi}^{(j)}(R') = \Gamma_{\phi}^{(j)}(R')$, then
$$(\Psi_{\varkappa\lambda}^{(ij)},\hat{O}\phi_{\mu\nu}^{(ij)}) = \sum_{\eta=1}^{\ell_i}\sum_{\xi=1}^{\ell_j}(\Psi_{\eta\xi}^{(ij)},\hat{O}\phi_{\eta\xi}^{(ij)})\frac{\delta_{\varkappa\mu}}{\ell_i}\frac{\delta_{\lambda\nu}}{\ell_j} .$$

Proof: The steps are exactly the same as those given for
theorem (ix). The one step which has not been made rigorous
is that one requiring the unitarity of the direct product
matrix. However it is easily proved, under the definition
of matrix multiplication for direct product matrices, that
the direct product of unitary matrices is a unitary matrix.
And so the proof is exactly the same.

   Theorem (x) may similarly be restated in the
language of direct product groups.

   These theorems are most frequently applied to inversion symmetry, where G' is the so-called parity group
{E,i} and a special set of notations is conventionally used
to distinguish the representations of direct products of
other groups with the parity group. The representation in
which E and i are represented by +1 and +1 is called Γ^+.
That in which E and i are represented by +1 and -1, respectively, is called Γ^-. Now if the representations of G are
called $\Gamma^{(i)}$, then those of the direct product of G with the
inversion group are called $\Gamma^{(i)+} \equiv \Gamma^{(i)} \times \Gamma^+$ and $\Gamma^{(i)-} =$
$\Gamma^{(i)} \times \Gamma^-$, and the wave vectors are called "even" if they
transform according to the first and "odd" if they transform
according to the second.

In this language, theorem (xii) states that the matrix element will be zero unless both functions are even or both are odd. If the operator itself is not invariant under the parity group, then the generalized version of theorem (x) would say that if $\hat{O}$ were odd, then the two functions would have to have opposite parity in order that the matrix element be non-zero, although if it were even, the functions would have to have the same parity.

It should perhaps be remarked that since the representations of the parity group are one-dimensional, the wave vectors which transform according to irreducible representations of the direct product group require no parity subscript. A typical labeling would be $\Psi_{\varkappa}^{(i)+}$, rather than $\Psi_{\varkappa\lambda}^{(i)+}$.

Let us now present the character table for the cubic group with inversion symmetry and the double-valued representations. This is the group O_h' with 96 elements. (O has 24 elements; the double group O' has 48; and inversion brings the number to 96.)

	$6\bar{S}_4$	$6S_4$	$8\bar{S}_6$	$8S_6$	$12\sigma_d$	$6\sigma_v$	$\bar{1}$	1	$6\bar{C}_4$	$6C_4$	$8\bar{C}_3$	$8C_3$	$12C_2$	$6C_4^2$	$\bar{E}$	E
Γ_1^+, A_1	1	1	1	1	1	1	1	1	1	1	1	1	1	1	1	1
Γ_2^+, A_2	-1	-1	1	1	-1	1	1	1	-1	-1	1	1	-1	1	1	1
Γ_3^+, E	0	0	-1	-1	0	2	2	2	0	0	-1	-1	0	2	2	2
Γ_4^+, T_1	1	1	0	0	-1	-1	3	3	1	1	0	0	-1	-1	3	3
Γ_5^+, T_2	-1	-1	0	0	1	-1	3	3	-1	-1	0	0	1	-1	3	3
Γ_6^+	$-\sqrt{2}$	$\sqrt{2}$	-1	1	0	0	-2	2	$-\sqrt{2}$	$\sqrt{2}$	-1	1	0	0	-2	2
Γ_7^+	$\sqrt{2}$	$-\sqrt{2}$	-1	1	0	0	-2	2	$\sqrt{2}$	$-\sqrt{2}$	-1	1	0	0	-2	2
Γ_8^+	0	0	1	-1	0	0	-4	4	0	0	1	-1	0	0	-4	4
Γ_1^-, A_1	-1	-1	-1	-1	-1	-1	-1	-1	1	1	1	1	1	1	1	1
Γ_2^-, A_2	1	1	-1	-1	1	-1	-1	-1	-1	-1	1	1	-1	1	1	1
Γ_3^-, E	0	0	1	1	0	-2	-2	-2	0	0	-1	-1	0	2	2	2
Γ_4^-, T_1	-1	-1	0	0	1	1	-3	-3	1	1	0	0	-1	-1	3	3
Γ_5^-, T_2	1	1	0	0	-1	1	-3	-3	-1	-1	0	0	1	-1	3	3
Γ_6^-	$\sqrt{2}$	$-\sqrt{2}$	1	-1	0	0	2	-2	$-\sqrt{2}$	$\sqrt{2}$	-1	1	0	0	-2	2
Γ_7^-	$-\sqrt{2}$	$\sqrt{2}$	1	-1	0	0	2	-2	$\sqrt{2}$	$-\sqrt{2}$	-1	1	0	0	-2	2
Γ_8^-	0	0	-1	1	0	0	4	-4	0	0	1	-1	0	0	-4	4

TRANSITION PROBABILITIES

We consider optical transitions between one-electron excited states of a hydrogen-like atom. From quantum mechanics we know that the transition rate is given by Fermi's Golden Rule:

$$W_{fi} = \frac{2\pi}{\hbar} \mid M_{fi} \mid^2 \rho_f \; \delta(E_f - E_i)$$

where M_{fi} is the matrix element of the interaction inducing the transition between initial and final states. If $M_{fi} = 0$, then the transition is forbidden, and we are led to specific selection rules. It is our purpose here to show that the symmetry properties of the Hamiltonian are responsible for most selection rules. We consider the four one-electron states of the $n = 4$ level: 4s, 4p, 4d, and 4f. We saw earlier how to determine the spin-orbit and cubic field splittings of these levels by group-theoretical methods. The result is the graph shown below.

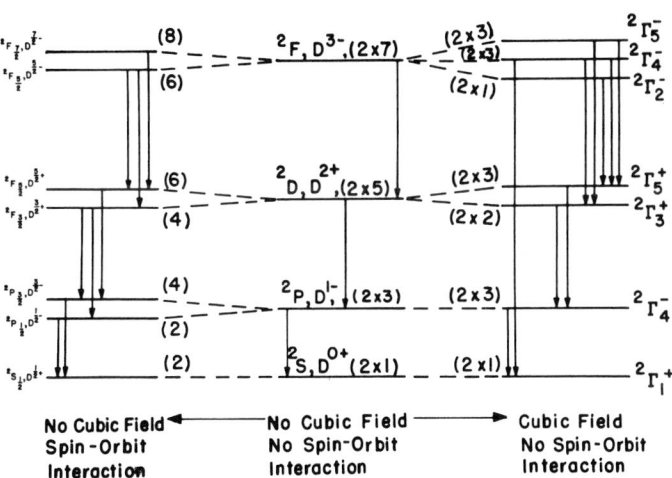

We have included the parity designations appropriate for the spherical harmonics describing the unperturbed state. The parity designations for the perturbed states follow automatically from the methods we have used.

We take the interaction to be that between the dipole moment of the atom and the electric field, which gives the largest component of electro-magnetic radiation. Thus the matrix element which determines the electric dipole selection rules, therefore, is $(\Psi_i, \underline{r}\Psi_f)$, since $e\hat{\underline{r}}$ is the dipole moment operator. For a magnetic dipole, we would choose $(\Psi_i, \hat{\underline{\mu}}\Psi_f) \propto (\Psi_i, \hat{\underline{L}}\Psi_f)$.

In order to apply the theorems developed above, we must find the representation of the symmetry group of the Schrödinger equation according to which the functions (x,y,z) transform. We do this for each of the three cases of our table.

Case (A): No spin-orbit or crystal field. The group of the Schrödinger equation is the full rotation group plus inversion applied independently in coordinate space and spin space. Thus it is a direct product group, and so by theorem (xii), page 120, we may investigate separately the transformations in spin space and in coordinate space.

Since rotations in spin space do not affect functions of coordinates alone, (x,y,z) must transform according to D^0 there. And so by theorem (xii), the matrix element will be zero unless $\Psi_i(\underline{r},\sigma_z)$ and $\Psi_f(\underline{r},\sigma_z)$ transform under rotations in spin space according to the same irreducible representation of the full rotation group. (Spinors are chosen to have even parity so that we may ignore inversion symmetry.) In other words, the selection rule is $\Delta S = 0$.

In addition we must have a selection rule coming from rotational symmetry in coordinate space. We know that the atomic p-functions, which transform according to D^1, can be written $xf(|r|)$, $yf(|r|)$, $zf(|r|)$. So (x,y,z) must transform as D^{1-}. (The momentum operator $\underline{p} \propto \underline{\nabla}$ also transforms according to D^{1-}, and the angular momentum operator $\hat{\underline{L}} = \underline{r}\times\underline{p}$ transforms according to D^{1+}.) To get a non-zero matrix

element, the representation according to which Ψ_i transforms and that according to which $\underline{r}\Psi_f$ transforms must have
at least one irreducible component in common. Since each
Ψ_i and each Ψ_f transforms according to an irreducible
representation of the group of the Schrödinger equation,
we must discover the conditions under which $\Gamma^{1-} \times \Gamma^{(f)}$ contains $\Gamma^{(i)}$ as an irreducible component.

Now $\Gamma^{1-} = \Gamma^1(\alpha) \times \Gamma^-(R)$, where α is a proper rotation
and R is either E or i. Hence $\Gamma^{1-} \times \Gamma^{(f)\pm} = \Gamma^1(\alpha) \times \Gamma^{(f)}(\alpha)$
$\times \Gamma^-(R) \times \Gamma^{\pm}(R)$. It is easy to confirm from the character
table of the parity group, page 48, that $\Gamma^-(R) \times \Gamma^{\pm}(R) =$
$\Gamma^{\mp}(R)$, while $\Gamma^+(R) \times \Gamma^{\pm}(R) = \Gamma^{\pm}(R)$. We conclude at once,
therefore, that $\Psi^{(i)}$ and $\Psi^{(f)}$ must have opposite parities:
$\Delta p = \pm 2$.

In addition, $\Gamma^1(\alpha) \times \Gamma^{(f)}(\alpha)$ must contain $\Gamma^{(i)}(\alpha)$,
according to theorem (xii). But we have proved on page 99
that $\Gamma^1(\iota) \times \Gamma^{(f)}(\alpha)$ contains the components $\Gamma^{(f-1)}(\alpha)$,
$\Gamma^{(f)}(\alpha)$, and $\Gamma^{(f+1)}(\alpha)$. Hence we get zero matrix elements
unless f = i + 1, i, or i - 1. In other words, $\Delta L = 0, \pm 1$.
If f = 0, then the only component of $\Gamma^{(i)}(\alpha) \times \Gamma^{(f)}(\alpha)$ is
$\Gamma^{(1)}(\alpha)$; so we get one exception to this rule which forbids
transitions between L = 0 states.

We can continue the analysis further by using the
second half of theorem (xii), page 120 and get selection
rules on ΔM, but we omit that part of the analysis here.

Case (B): Spin-orbit effects, no crystal field. The
group of the Schrödinger equation is now the full rotation
group plus inversion, applied simultaneously in spin-space
and coordinate space. The eigenfunctions are now labeled
according to the total angular momentum J and transform
according to D^J under this group. An analysis exactly
like that above gives the selection rule $\Delta J = 0, \pm 1$, but
$J = 0 \nrightarrow J = 0$, and similar rules for ΔM_J. The parity selection rule survives the loss of symmetry, but the others
concerning ΔL and ΔS become only approximations, valid when
spin-orbit coupling is very weak.

Case (C): No spin-orbit effects, but cubic field. The
group of the Schrödinger equation is now the full rotation

group in spin space and the cubic group with inversion O_h in coordinate space. As in case (A), the spin-space symmetry leads to the selection rule $\Delta S = 0$. And, as in that case, this is automatically satisfied since every level is a doublet $(S = \frac{1}{2})$ level.

To get the selection rules from the coordinate space symmetry, we must first find the representation of O_h according to which the functions (x, y, z) transform. Examples of this method are shown on pages 58-69. It can be easily verified that (x, y, z) transform according to Γ_4 of O. Clearly they have negative parity under the inversion group. Hence under O_h they must transform as $\Gamma_4 \times \Gamma^- = \Gamma_4^-$.

This negative inversion symmetry tells us, as in case (A), that the matrix element will be zero unless the initial and the final states have opposite parities. This selection rule will evidently remain in all systems with inversion symmetry. The rule for magnetic dipole interaction states that the two parities must be the same, since the parity of the angular momentum operator is positive.

We get the selection rule analogous to $\Delta L = 0, \pm 1$ in case (A) by applying theorem (x) once more. $\Gamma_4 \times \Gamma_f$ must contain Γ_i or else M_{fi} will be zero. Using the character table on page 122 we can confirm the following reductions

$$\text{(i)} \quad \Gamma_4 \times \Gamma_1 = \Gamma_4$$
$$\text{(ii)} \quad \Gamma_4 \times \Gamma_2 = \Gamma_5$$
$$\text{(iii)} \quad \Gamma_4 \times \Gamma_3 = \Gamma_4 + \Gamma_5$$
$$\text{(iv)} \quad \Gamma_4 \times \Gamma_4 = \Gamma_1 + \Gamma_3 + \Gamma_4 + \Gamma_5$$
$$\text{(v)} \quad \Gamma_4 \times \Gamma_5 = \Gamma_2 + \Gamma_3 + \Gamma_4 + \Gamma_5$$

We actually need only the characters for the group O, that is, the first five rows of the table on page 122 and the first five columns excluding the barred classes.

Now let us look at the term diagram on page 123. If

the final state transforms according to Γ_1^+, then line (i) and the parity selection rule indicate that the initial state must be a Γ_4^-. Similarly the Γ_4^- state may be connected only with Γ_1^+, Γ_3^+, Γ_4^+, or Γ_5^+. Thus the five equations above contain all selection rules of the cubic group O.

We can now apply the second half of theorem (xii) to get additional selection rules analogous to those on ΔM in case (A).

The term diagram on page 123 contains all the allowed electric dipole transitions between the n = 4 levels, as is easily confirmed by application of the selection rules derived above. The spectra to be expected are shown in the figure below.

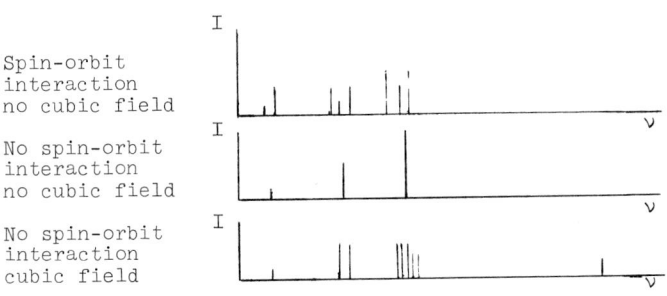

Spin-orbit interaction no cubic field

No spin-orbit interaction no cubic field

No spin-orbit interaction cubic field

We see that a loss of symmetry is accompanied not only by a splitting of degenerate levels but also by a breaking down of selection rules for transitions between levels. If there was no symmetry there would be no degeneracy, and all transitions would be allowed. This follows from theorem (x), where the group is the identity element, whose single representation is $\Gamma_1 = 1$.

The intensities of the spectral lines can be inferred from the second half of theorem (x) or (xii). The significant result there is that all the non-zero matrix elements taken between eigenfunctions which transform according to the same irreducible representation are _identical_. Each one, according to the theorem, is the average of all. Hence the total transition rate between two levels for which a transition is allowed will be proportional to one of these matrix elements and to the number of such elements - that is, the degeneracy of the final state.

Although we shall leave it at this point, it is easy to continue this example by adding spin-orbit effects to case (C). We have already seen how to get the splittings in this case (pp. 111-12), and we get the new selection rules from theorem (x) in the usual way. The group of the Schrödinger equation in this case is the cubic group, with inversion, applied simultaneously in spin-space and coordinate space. We must use the double-valued representations of O_h now because the spin in our example is $\frac{1}{2}$. We must also discover how spin functions, which transform as $D^{\frac{1}{2}+}$ under the full group, transform under O_h.

Similarly, we might inquire into the effects on selection rules of further loss of symmetry induced, for example, by straining along some symmetry axis of O_h.

Before we turn to perturbation theory, there is one general theorem on selection rules which we may easily prove.

Theorem: The only allowed optical transitions between states whose eigenfunctions have the full symmetry of the Hamiltonian are those arising from multipole modes which also have the full symmetry of the Hamiltonian.

Proof: By hypothesis both eigenfunctions transform according to the identical representation Γ^1. If a given multipole mode transforms according to $\Gamma^{(m)}$, theorem (x), page 117, requires that $\Gamma^1 x \Gamma^{(m)}$ contains Γ^1, and this is only possible if $\Gamma^{(m)} = \Gamma^1$.

Since there is no monopole radiation it is necessary, for Hamiltonians with high symmetry, to go to high order multipoles to find a mode which transforms according to Γ^1. For the full rotation group such mode does not exist.

We have seen special cases of this theorem where we found no electric dipole transitions between states with $L = 0$ or with $J = 0$ in the free atom. We also have found no matrix element connecting Γ_1 with Γ_1 in the case of cubic symmetry.

PERTURBATION THEORY

The problem in perturbation theory is similar to that in the theory of transition rates, but with a few essential differences. For our example let us formally set up a calculation scheme to determine the amount of splitting of degeneracies when a system whose energy levels and eigenfunctions we know is placed in a potential of known form. Specifically, let the unperturbed system be the free atom we considered in the transition rate problem, but we will ignore spin completely. We furthermore suppose that the $n = 4$ shell is far enough removed in energy from the remaining levels that the energy denominators of the higher order perturbation terms connecting with other levels will be large enough to allow us to ignore their presence. As usual we write $H = H_0 + V$, form the matrix of H using the eigenfunctions of H_0, and systematically diagonalize the matrix according to the rules of perturbation theory.

Case (A): $V = x^2y^2 + y^2z^2 + z^2x^2$. The group of H_0 is the full rotation group with inversion. The group of maximum symmetry of V, and hence of H, is O_h, as may be found by applying successively all its operations to V. We have already seen on page 123, the qualitative features of the splitting which results.

There are 16 eigenfunctions of the $n = 4$ shell, and so the calculation of the new energy levels involves the solution of a 16 x 16 secular equation, that is, the calculation of 256 matrix elements of V as well as the solution of a 16 x 16 determinant and a $16^{\underline{th}}$ order algebraic equation. As we shall see, however, most matrix elements of V, and hence of H, are zero when V has a high degree of symmetry.

Now $H_{ij} \equiv (\Psi_i, \hat{H}\Psi_j) = H_{0,ij} + V_{ij} = E_{0i}\delta_{ij} + V_{ij}$. Hence the off-diagonal elements will be zero if, and only if, $V_{ij} = 0$. We shall apply our matrix element theorems to $V_{ij} = (\Psi_i, V\Psi_j)$, but we must decide whether the group of

$\hat{H}_0$ or the group of $\hat{H}$ is better for our problem. Since the final states are to be labeled according to the group of $\hat{H}$, this would appear to be a good choice more often than not. But there is no reason a priori for using the symmetry of $\hat{H}_0$. The methods will be somewhat different in the two cases. Here we use the group of $\hat{H}$, O_h.

O_h is a direct product group, and V is an invariant operator; so we use theorem (xii), page 120 . On page 123 we have labeled the levels of any Hamiltonian with the symmetry of O_h according to the irreducible representations to which its eigenfunctions transform. Theorem (xii) tells us at once that there will be no non-diagonal matrix elements of V except those taken between the eigenfunctions of the two Γ_4^- levels. In addition, all diagonal elements may be non-zero.

However, we are not using eigenfunctions of $\hat{H}$, but only those of $\hat{H}_0$. Nevertheless we know that under the operations of O_h the eigenfunctions of $\hat{H}_0$ will transform according to a (generally reducible) representation of O_h. Hence the matrix elements taken with respect to the eigenfunctions of $\hat{H}_0$ will be zero unless the representations they generate of the group of $\hat{H}$ contain an irreducible component in common. We see that in our example only the f- and p-functions have this property.

$$(\Psi_f, V\Psi_d)=(\Psi_f, V\Psi_s)=(\Psi_d, V\Psi_p)=(\Psi_d, V\Psi_s)=(\Psi_p, V\Psi_s)=0.$$

This is already a great simplification, since the Hamiltonian matrix may be written in reduced, or block form. An elementary theorem of matrix algebra tells us that the determinant of a matrix in block form is the product of the determinants of the blocks; so each block may be diagonalized independently of the others.

The block coming from the s-function is one-dimensional and so diagonal. The d-block is 5-dimensional. And the remaining pf-block is 10-dimensional. Let us look more closely at the d-block. We know from the graph on page 123 that the 5 d-functions generate a representation of O_h which may be reduced into Γ_3^+ and Γ_5^+. By means of the projection operators we may find the three independent linear combinations of d-functions which transform according to Γ_5^+ and the two others which transform according to Γ_3^+.

If we use these particular d-functions, then the second part of theorem (x) tells us that all off-diagonal elements within the d-block will be zero and that three of the diagonal elements will be identical with one another, as will two others.

Next we follow the same procedure with the 7 f-functions, and we get one function transforming according to Γ_2^-, three according to Γ_4^-, and three more according to Γ_5^-. Theorem (xii) tells us that the p-functions which transform according to Γ_4^- will give non-zero matrix elements of V only when taken with those f-functions which transform also according to Γ_4^-. It also tells us that the three sets of f-functions also do not mix and, furthermore, that within any one set the only non-zero elements are the diagonal ones that are equal to one another.

Finally we pick a linear combination of the three p-functions which transforms according to the same representation as the three Γ_4^- f-functions. Theorem (xii) tells us once more that the only non-zero matrix elements of V between these p-functions and these f-functions will be those taken between corresponding functions - the first p with the first f, the second p with the second f, and the third p-function with the third f-function - and that furthermore these three non-zero elements will be identical.

In summary there are only $16 + 2 \cdot 3 = 22$ non-zero matrix elements in the Hamiltonian matrix, and of these only 8 are distinct. Instead of having ·to do $\frac{256-16}{2} + 16 = 136$ integrals we need do only 8 in order to get the matrix of V. We finish the problem by diagonalizing one of the three equivalent 2 x 2 matrices arising from the last step above.

Case (B): V = xyz. Let us suppose that now we add this new potential to the previous problem considered. That is, let $\hat{H}_0 = T + V_0(|\underline{r}|) + a(x^2y^2+y^2z^2+z^2x^2)$, $\dot{H} = \hat{H}_0 + V$. V does not have the symmetry of $\hat{H}_0$, which is O_h, but has instead the symmetry of the tetrahedron T. In this example, however, we shall work with the group of $\hat{H}_0$ instead of that of $\hat{H}$.

First we must discover the irreducible representations according to which V transforms under O_h. It is easy to confirm that the answer is Γ_2^-. According to theorem (x), page 117, suitably generalized to handle representations of direct product groups, we must form the direct product representation $\Gamma_2^- \times \Gamma_j^+$ for all Γ_j^+ of the unperturbed state:

$$\Gamma_2^- \times \Gamma_1^+ = \Gamma_2^-$$
$$\Gamma_2^- \times \Gamma_4^- = \Gamma_5^+$$
$$\Gamma_2^- \times \Gamma_3^+ = \Gamma_3^-$$
$$\Gamma_2^- \times \Gamma_5^+ = \Gamma_4^-$$
$$\Gamma_2^- \times \Gamma_2^- = \Gamma_1^+$$

The only non-zero matrix elements will be those taken between the corresponding states in the columns indicated by arrows. Arguing exactly as before we see that there will be a 2 x 2 matrix from the Γ_2^- and Γ_1^+ states, a 9 x 9 matrix from the six Γ_4^- states and the three Γ_5^+ states, a 3 x 3 diagonal matrix from the three Γ_5^- states, and a 2 x 2 diagonal matrix from the two Γ_3^+ states. To go further than this we work with the representations of T, the group of $\hat{H}$, as well as the representations of O_h. From the character tables of pages 67 and 122 it is easily seen that the inclusion of V = xyz term removes some symmetry and changes

$$\text{Representations of } O_h \begin{cases} \Gamma_1^+ , \ \Gamma_2^+ \ \rightarrow \ A \\[2mm] \Gamma_3^+ \ \rightarrow \ E \\[2mm] \Gamma_4^+ , \ \Gamma_5^+ \ \rightarrow \ T \end{cases} \text{Representations of } T$$

Our problem is now reduced to calculate

$\Gamma_1^+ + \Gamma_2^- \ \rightarrow \ 2A$: One matrix element and a 2 x 2 secular equation

$2\Gamma_4^- + \Gamma_5^+ \ \rightarrow \ 3T$: Two matrix elements and one (three-equivalent) 3 x 3 secular equation.

The levels $\Gamma_3^+ \rightarrow E$ and $\Gamma_5^- \rightarrow T$ are unaffected by the perturbation.

DIRECTED VALENCE AND EQUIVALENT FUNCTIONS

We conclude this chapter with a discussion of a group-theoretical approach to the problem of constructing from the complete set of free-atom orbitals another complete set which displays explicitly the known symmetry of molecules or crystals in which the atom is bonded in some geometrical arrangement with its neighbors. The free-atom orbitals are eigenfunctions of a central potential problem; therefore they transform according to irreducible representations of the full rotation group, whereas the exact eigenfunctions of the molecule, for example, must transform according to irreducible representations of a group of lower symmetry. In the section on perturbation theory we saw the virtue of selecting our approximate wave functions so that they have the symmetry of the final state. Our purpose here is to show how to do this.

From the theory of valence bonding we know that the major contribution to the binding energy comes from the

so-called <u>exchange energy</u>, the size of which is determined
by the overlap integral of the charge clouds of neighboring
atoms. Consequently those orbitals which concentrate the
probability distribution along preferred directions are
most apt to lead to bonding with other similar distribu-
tions on neighboring atoms. These are the so-called bonding orbitals. The complete set of orbitals will also include orbitals with nodes along the directions of the
bonding orbitals. These are called "anti-bonding orbitals"
for obvious reasons.

Consider the carbon atom whose free-atom configuration is $1s^2 2s^2 2p^2$ in the ground state. We know that in
the methane series of molecules, $C_n H_{2n+2}$, and in the dia-
mond crystal lattice we have a characteristic tetrahedral
bonding. Can we form suitable approximate wave functions
which display this symmetry from the low-lying atomic
orbitals? If we could form four different linear combinations of four different atomic orbitals such that the
charge distribution in each new orbital lay along one of
the trigonal axes of a tetrahedron, we could use them in
this application.

The essential feature of such orbitals is that they
transform into one another under all operations of the
group of the molecule with, at most, only a change of phase.
Such functions are called "equivalent," and by definition
they generate a representation of the group, although not
usually an irreducible one.

We show here how to use the methods of group theory
to find equivalent functions. In the simplest case we
imagine that we have a scalar quantity, which we shall
symbolize by a point, attached to the ends of the four
tetrahedral axes. We number these equivalent points as
shown in the figure on page 67 . Now we generate a representation of the group T_d by finding the matrices which
represent the corresponding permutations of the numbered
points. For example, the C_3 rotation about the (111) axis
corresponds to the following permutation: $p_1 \rightarrow p_1$, $p_2 \rightarrow p_3$,
$p_3 \rightarrow p_4$, $p_4 \rightarrow p_2$. The permutation matrix would be

$$\begin{pmatrix} 1 & 0 & 0 & 0 \\ 0 & 0 & 0 & 1 \\ 0 & 1 & 0 & 0 \\ 0 & 0 & 1 & 0 \end{pmatrix}$$

We shall need only the characters of this representation, and it is easy to see that the character is always equal to the number of points left unaffected by the transformation. The character table for T_d below shows the characters of this new representation and typical functions which transform according to the representations.

		E	$8C_3$	$3C_2$	$6\sigma_d$	$6S_4$
(r)	A_1	1	1	1	1	1
	A_2	1	1	1	-1	-1
$(x^2-y^2,\ 3z^2-r^2)$	E	2	-1	2	0	0
	T_1	3	0	-1	-1	1
$(x,y,z);(xy,yz,zx)$	T_2	3	0	-1	1	-1
	Γ_σ	4	1	0	2	0
	Γ_π	12	0	0	2	0

It is obvious that the new representation is reducible into $A_1 + T_2$, and it is easy to confirm that its characters are these we have labeled Γ_σ. We know therefore that any A_1 function and any set of three T_2 functions may be combined into four equivalent functions. In view of the atomic configuration of carbon we are encouraged to try an s-function and three p-functions. If we normalize these so that $\int p_x^2 d\Omega = 4\pi$, and so forth, then the combination $p_x + p_y + p_z$ transforms under T_d into $p_x - p_y - p_z$, $-p_x - p_y + p_z$ and $-p_x + p_y - p_z$.

These four functions are not linearly independent, but by
adding A.s to each we make them so. To make the resulting
functions ortho-normalized, we take A = 1 and multiply
the whole by $\frac{1}{2}$.

$$u_{111} = \tfrac{1}{2}(s + p_x + p_y + p_z)$$

$$u_{\bar{1}\bar{1}1} = \tfrac{1}{2}(s - p_x - p_y - p_z)$$

$$u_{1\bar{1}\bar{1}} = \tfrac{1}{2}(s + p_x - p_y - p_z)$$

$$u_{\bar{1}1\bar{1}} = \tfrac{1}{2}(s - p_x + p_y - p_z)$$

In terms of the usual polar angles we can write $u_{111} =$
$f(r)\tfrac{1}{2}(1 + \sqrt{3} \sin\theta(\cos\varphi + \sin\varphi) + \sqrt{3} \cos\theta)$. Now along the
positive (111) direction $\cos\theta = \frac{1}{\sqrt{3}}$, $\sin\theta = \frac{\sqrt{2}}{3}$, $\cos\varphi =$
$\sin\varphi = \frac{1}{\sqrt{2}}$, and $u_{111}(111) = 2f(r)$. In the opposite direction $\bar{1}\bar{1}\bar{1}$ the $\cos\theta$, $\cos\varphi$, and $\sin\varphi$ change signs and
$u_{111}(\bar{1}\bar{1}\bar{1}) = -1f(r)$. $u_{111} = 0$ along $(1\bar{1}\bar{1})$, $(\bar{1}1\bar{1})$, $(\bar{1}\bar{1}1)$,
where its equivalent functions are maxima. Finally along
$(\bar{1}11)$, $(1\bar{1}1)$ and $11\bar{1}$, $u_{111} = +1f(r)$. Thus u_{111} has one
large lobe along 111, has four smaller ones making tetra-
hedral angles with one another but lying in opposite directions from the large lobes of the equivalent functions, and
has nodes along the large lobes of the other three equivalent functions.

Note that to reach this configuration one of the
s-electrons has to go up into a p-level, but the gain in
binding energy should compensate for this. Another set
of equivalent functions of this type would come from an
s-function and the three d-functions d_{xy}, d_{yz}, d_{zx}, but
this would be energetically unfavorable, since one
would have to go all the way to the n = 3 shell to get
d-orbitals. Nevertheless these higher-lying orbitals do,
in principle, contribute a small share to the exact eigenfunctions.

But there is a wholly different class of tetrahe-
drally equivalent functions, which we show how to find now.
Suppose that instead of associating a scalar quantity with
the corners of our tetrahedron, we place a vector quantity
there. We represent it by a vector pointing from the
corner along one of the edges towards another corner. It
is not possible to place one such vector quantity on each
corner, and the symmetry suggests putting three on each
corner along the three edges - a total of twelve such
quantities.

Now we associate a number with each of these twelve
little vectors and, as before, work out the traces of the
permutation matrices corresponding to each operation of T_d.
That is, we count the number of vectors which go into themselves under the operation. If a vector goes into the
negative of itself we subtract its contribution to the
trace. (This does not occur in our example.)

The result is shown as Γ_π in the table on page 135.
By applying theorem VIII, page 42 , we find that $\Gamma_\pi = A_1 +
E + T_1 + 2T_2$. Since $A_1 + T_2 = \Gamma_\sigma$, we may view this as a
linear combination of the four σ-type orbitals found above
with eight others yet to be found. We shall not pursue
this example further except to note that the symmetry is
similar to that of the σ-orbitals with major lobes along
the tetrahedral axes. But when we look along one of the
axes one of the twelve equivalent functions will have lobes,
as shown in the diagram in solid lines, in a direction at
right angles to the tetrahedral axes.

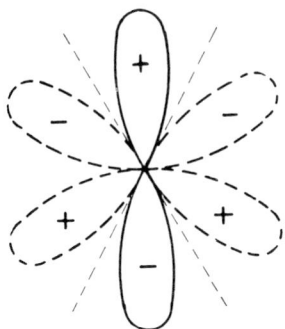

Two others will have a lobe structure as shown in dashed lines in the figure - thus three vector-like functions per axis.

These π-orbitals evidently cannot be constructed from s-, p-, and d-functions alone, since none transform as T_1; therefore they do not contribute very much to the bonding of carbon.

Since equivalent functions do not transform according to an irreducible representation of the symmetry group of the molecule, they cannot be eigenfunctions of the system. But recall that the approximate wave functions are to be formed as a _product_ of, for example, the four σ-functions, and this _product_ function _does_ transform as an irreducible representation - the identical representation, in fact, of T_d. And so it is a very appropriate choice for an approximate wave function.

In graphite and the benzene-ring compounds the bond angle is 120° and the major bonds are co-planar. The symmetry group is D_{3h}. It is possible, by the means described above, to form σ- and π-type orbitals of this symmetry. We shall not do it here, however, but only point out that the σ-orbitals lying lowest are sp^2 "hybrids," as they are called, and there are two sets of π-orbitals. One of these sets, the pd^2 hybrids, is responsible for the double bonding of certain atoms.

VI

Applications to Solid State Physics

For the atomic physicist the model atom has, in first approximation, a perfectly spherically symmetric Hamiltonian whose symmetry is broken in higher approximations by residual electrostatic and spin interactions. For the solid state physicist the model solid has, in first approximation, a Hamiltonian with perfect translational symmetry, which symmetry is broken in higher approximation by effects such as vibrations of the lattice, by impurity effects, by surface effects, and others.

To say that the Hamiltonian has perfect translational symmetry means that there exists a specific set of vectors, $\underline{a}_1$, $\underline{a}_2$, and $\underline{a}_3$, called "primitive vectors", such that the form of the Hamiltonian is left invariant when it is expressed as a functional of $\underline{x} + \underline{t}$ instead of $\underline{x}$, where $\underline{t} = n_1 \underline{a}_1 + n_2 \underline{a}_2 + n_3 \underline{a}_3$ and n_1, n_2, n_3 are <u>any</u> set of integers. In other words, if an electron at any point $\underline{x}$ were moved to $\underline{x} + \underline{t}$ it would be subject to exactly the same forces. To put it in a different form, once the Hamiltonian is defined in a specific primitive cell of space, it is determined everywhere by performing all of the translations $\underline{t}$.

The parallelepiped whose edges are $\underline{a}_1$, $\underline{a}_2$, and $\underline{a}_3$ is such a primitive cell; but any other portion of space defined by dividing that parallelepiped arbitrarily into any number of pieces and subjecting each piece to a generally different translation $\underline{t}$ also is a **primitive** cell. Thus there is some arbitrariness in the definition of the primitive cell. There is also an arbitrariness in the choice of primitive vectors. For if $\underline{a}_1$, $\underline{a}_2$, and $\underline{a}_3$ are primitive, so is $\underline{a}_1$, $\underline{a}_2$, $\underline{a}_3 + 26\underline{a}_1 + 37\underline{a}_2$, since the old translation $\underline{a}_3$ can be expressed as $\underline{t} = -26\underline{a}_1 - 37\underline{a}_2 + (\underline{a}_3 + 26\underline{a}_1 + 37\underline{a}_2)$. Every primitive cell however must have the same volume.

If we arbitrarily choose a point in space, then the set of all points equivalent to that point by translational symmetry is called a lattice. If we choose a new origin displaced an amount $\underline{\tau}$ from the old one, then we get a new lattice of the same structure as before but displaced an amount $\underline{\tau}$. If $\underline{\tau} = \underline{t}_{n_1 n_2 n_3}$, one of the lattice translations, then the new lattice is identical with the old one.

One useful primitive cell is the so-called Wigner-Seitz cell. From an arbitrarily chosen origin we draw the lattice vectors to all the nearby equivalent points. Then we construct planes through the midpoints of and perpendicular to these vectors. The smallest volume about that origin which is enclosed by these intersecting planes is the Wigner-Seitz cell. This choice of primitive cell is convenient because it displays in a rather obvious manner any rotational symmetries possessed by the lattice. The choice of origin, which is arbitrary as far as translational symmetry is concerned, will be determined by other considerations. If, for example, the Hamiltonian is that of a periodic arrangement of point ions, then the most likely candidate for the origin is a point occupied by an ion.

Earlier we enumerated the thirty-two point groups consistent with lattice translational symmetry. Lattices may be classified according to the point groups of rotations and reflections performed about a lattice point, which leave the lattice of equivalent points invariant. The most general lattice, for example, is invariant under the groups C_1 and S_2. But to be invariant under C_2 we must put restrictions on the lattice. Thus if a set of primitive vectors can be found for which $\underline{a}_1 \cdot \underline{a}_2 = \underline{a}_1 \cdot \underline{a}_3 = 0$, it will be invariant under C_2. Such a lattice will also be invariant under C_{1h} and C_{2h}. We can continue in this

way until we exhaust the thirty-two groups, and we find
fourteen distinct lattices - the "Bravais lattices". It
may happen that the same set of point groups leaves more
than one Bravais lattice invariant. Thus three distinct
Bravais lattices are left invariant by all five groups,
T, T_h, T_d, O, and O_h. These are the familiar simple cubic,
face-centered cubic, and body-centered cubic lattices
(sc, fcc, and bcc). Thus we may further classify the
Bravais lattices into the so-called "crystal systems", defined on page 92 . We shall not describe the Bravais
lattices here because they are widely available in the
literature.

This does not, however, complete a description of
the full symmetries possible for an object with translational symmetry. Suppose that at each lattice point we
place an object with the symmetry of some particular point
group, not necessarily one of the crystallographic point
groups. If, for example, we place an object with full rotational symmetry at each point of the lattice, the complete group of this structure will be the point group of
maximum symmetry of the particular Bravais lattice and
the translation group of that lattice. If, on the other
hand, we place there an object whose point group symmetry
is C_1, then the complete group of this structure is merely
the translation group, even though the Bravais lattice
might itself have high point group symmetry.

A systematic enumeration of all symmetries of
periodic arrangements of entities distributed around
lattice points yields 230 distinct groups, the so-called
"space groups". They involve, in addition to the point
group operations - inversion, rotations, and reflections -
and translations, new elements which are combinations of
rotations and translations - screw axes - or combinations
of reflections and translations - glide planes.

We now introduce a systematic notation for all operations involved. A general space-group operation is de-
noted by $\{\check{R} \mid \underline{\tau}\}$, where R is a point-group operation and $\underline{\tau}$

a translation, not necessarily a member of the translation group. Operationally $\{R \mid \underline{\tau}\}$ is defined by

$$\{R \mid \underline{\tau}\} \; \underline{x} = R\underline{x} - \underline{\tau}$$

The rule of combination, that is, the group multiplication, can be easily derived:

$$\{S \mid \underline{\tau}_2\}\{R \mid \underline{\tau}_1\} \; \underline{x} = \{S \mid \underline{\tau}_2\} \; (R\underline{x} - \underline{\tau}_1) = SR\underline{x} - S\underline{\tau}_1 - \underline{\tau}_2$$

$$\therefore \qquad \{S \mid \underline{\tau}_2\} \; \{R \mid \underline{\tau}_1\} = \{SR \mid S\underline{\tau}_1 + \underline{\tau}_2\}.$$

The identity element is $\{E \mid 0\}$ and the inverse is

$$\{R \mid \underline{\tau}\}^{-1} = \{R^{-1} \mid -R^{-1}\underline{\tau}\}.$$

The space group $\mathcal{G}$ consists of all operations $\{R \mid \underline{\tau}\}$ which leave a given lattice invariant. The point group $\mathcal{R}$ connected with $\mathcal{G}$ is obtained by putting $\underline{\tau} = 0$ in all elements of $\mathcal{G}$. If, with a suitable choice of the origin in the lattice, <u>all</u> elements of $\mathcal{R}$ are elements of $\mathcal{G}$, then $\mathcal{G}$ is called a symmorphic group (there are only 73 symmorphic groups in the 230 space groups); otherwise, $\mathcal{G}$ is called non-symmorphic.

All the elements of $\mathcal{G}$ that are of the form $\{E \mid \underline{t}\}$ constitute the translation group, indicated by $\mathcal{J}$. In this case, $\underline{t}$ is necessarily a translation vector of the form $n_1\underline{a}_1 + n_2\underline{a}_2 + n_3\underline{a}_3$. $\mathcal{J}$ is an invariant subgroup of $\mathcal{G}$:

$$\{R \mid \underline{\tau}\}\{E \mid \underline{t}\}\{R \mid \underline{\tau}\}^{-1} = \{R \mid \underline{\tau}\}\{E \mid \underline{t}\}\{R^{-1} \mid -R^{-1}\underline{\tau}\} =$$

$$\{R \mid \underline{\tau}\} \; \{R^{-1} \mid -R^{-1} \; \underline{\tau} + \underline{t}\} = \{E \mid -RR^{-1} \; \underline{\tau} + R\underline{t} + \underline{\tau}\} = \{E \mid R\underline{t}\}$$

But $R\underline{t}$ is another translation vector and $\{E \mid R\underline{t}\}$ belongs to $\mathcal{J}$.

Even in the case of the symmorphic groups, where all $\underline{\tau}$ are equal to $\underline{t}$, $\mathcal{G}$ is <u>not</u> a direct product of a

translation group and a point group, as can be seen by
the non-commutativity

$$\{E \mid \underline{t}\}\{R \mid 0\}=\{R \mid \underline{t}\}; \quad \{R \mid 0\}\{E \mid \underline{t}\}=\{R \mid R\underline{t}\}$$

But since $\mathfrak{J}$ is an invariant subgroup, it is of interest
to study the factor group $\mathfrak{G}/\mathfrak{J}$. For symmorphic groups the
right (or left) cosets are given by

$$C_R = \Big\{\{E \mid \underline{t}\}\{R \mid 0\}\Big\} \quad = \quad \Big\{\{R \mid \underline{t}\}\Big\}$$

where the outer bracket denotes the entire complex of
elements formed by giving $\underline{t}$ all the translation vector
values. Under the multiplication rule for complexes

$$C_R C_S = \Big\{\{R \mid \underline{t}_1\}\{S \mid \underline{t}_2\}\Big\} = \Big\{\{RS \mid R\underline{t}_2+\underline{t}_1\}\Big\}= \Big\{\{T \mid \underline{t}_3\}\Big\}=C_T$$

where RS = T and $\underline{t}_3 = R\underline{t}_2 + \underline{t}_1$ is by necessity a translation vector, and when $\underline{t}_1$ and $\underline{t}_2$ range over all values,
$\underline{t}_3$ also runs on all values.

These considerations prove that for symmorphic
groups, the factor group $\mathfrak{G}/\mathfrak{J}$ is isomorphic with the point
group. Similar proof can be found for non-symmorphic
groups. Since irreducible representations of the factor
group are also irreducible representations of the group
itself, we have proved that in this case the irreducible
representations of $\mathfrak{R}$ are also irreducible representations
of the entire space group $\mathfrak{G}$. We might anticipate, therefore, that _all_ of the irreducible representations of any
space group of this simple kind may be compounded in some
simple way out of the irreducible representations of the
appropriate point group of rotations and the irreducible
representations of the translation group. This in fact is
the case, although the proof is very lengthy. G. F. Kos-
ter's article in _Solid State Physics_, Vol. 5, edited by
F. Seitz and D. Turnbull has an excellent presentation of
this proof. Our treatment will be less rigorous and less

complete, for our purpose is to display general results
in specific model examples.

<u>The Translation Subgroup $\mathfrak{J}$.</u>

Since this subgroup is common to all space groups
we shall study it first. A general element is the trans-
lation $\{E \mid \underline{t}\}$, where $\underline{t} = n_1\underline{a}_1 + n_2\underline{a}_2 + n_3\underline{a}_3$, such that
$\{E \mid \underline{t}\}\underline{x} = \underline{x} - \underline{t}$. The group of operators $\hat{P}_{\{E \mid \underline{t}\}}$, isomorphic with $\mathfrak{J}$, is defined in the usual way:

$$\hat{P}_{\{E \mid \underline{t}\}}\psi(\underline{x}) = \psi(\{E \mid \underline{t}\}^{-1}\underline{x}) = \psi(\underline{x} + \underline{t}).$$

The general element may also be written

$$\{E \mid \underline{t}\} = \{E \mid \underline{t}_1\}\{E \mid \underline{t}_2\}\{E \mid \underline{t}_3\}$$

where $\underline{t}_1 = n_1\underline{a}_1$, $\underline{t}_2 = n_2\underline{a}_2$, $\underline{t}_3 = n_3\underline{a}_3$. Since $\{E \mid \underline{t}_1\}$,
$\{E \mid \underline{t}_2\}$, and $\{E \mid \underline{t}_3\}$ commute with one another, we see
that $\mathfrak{J}$ is a direct product of three commuting subgroups
of the same type. Each of these subgroups is a cyclic
group whose representations we find on page 77, provided
that we require the group to close after a finite number
of repetitions of a given translation. Real, physical
crystals, of course, come to an end; so if we wish to
apply the theory of symmetry to them, we must make a reasonable approximation at the boundaries. One alternative
is to assume the crystal to be infinite in fact and to
employ the theory of the infinite, but still discontinuous,
translation group. Another alternative, which permits
us to work with finite groups, is the assumption that
each translation which would move a point inside the
crystal to another point outside, passing through one
face of the crystal, in fact moves back inside through
the opposite face, going into the crystal the same distance and in the same direction as the original vector
moved out. Thus if $\{E \mid \underline{t}_1\}$ is inside the crystal bound-
aries, we put $\{E \mid \underline{t}_1 + m\underline{L}_1\} = \{E \mid \underline{t}_1\}$ for all integral
m where L_1 is the distance between opposite boundaries.

This latter approximation is called the application of "periodic boundary conditions," and it leads to the same results as the infinite crystal approximation in the limit that L_1, L_2, $L_3 \rightarrow \infty$. We shall work with periodic boundary conditions, bearing in mind that it is only an approximation to the real crystal.

On page 78 we saw that the $r\underline{\text{th}}$ irreducible representation of the $N\underline{\text{th}}$ order group of translations in one dimension could be labeled by the quantity $k_1 = 2\pi r_1 / L_1$ and be given by the equation

$$\Gamma^{k_1}(n_1 a_1) = e^{ik_1 n_1 a_1}$$

Now we apply the theorem on page 72 to the direct product group $\mathfrak{J}$. Since matrix multiplication is ordinary multiplication (all representations are one-dimensional) we have the result that

$$\Gamma^{k_1}(n_1 a_1)\ \Gamma^{k_2}(n_2 a_2)\ \Gamma^{k_3}(n_3 a_3) = e^{i(k_1 n_1 a_1 + k_2 n_2 a_2 + k_3 n_3 a_3)}$$

is a representation of $\mathfrak{J}$, which we label

$$\Gamma^{k_1 k_2 k_3}(n_1 \underline{a}_1 + n_2 \underline{a}_2 + n_3 \underline{a}_3) \quad .$$

The theorems on page 72-3 also guarantee that these representations are irreducible, unique, and exhaust the list of irreducible representations of $\mathfrak{J}$.

These results immediately suggest a way of relabeling the representations. We think of (k_1, k_2, k_3) as a vector $\underline{k}$ specified with respect to three basis vectors $(\underline{b}_1, \underline{b}_2, \underline{b}_3)$ by

$$\underline{k} = \frac{2\pi r_1}{N_1} \underline{b}_1 + \frac{2\pi r_2}{N_2} \underline{b}_2 + \frac{2\pi r_3}{N_3} \underline{b}_3 = \underline{k}_1 + \underline{k}_2 + \underline{k}_3$$

and such that the scalar product with the translation

$$\underline{t} = n_1 \underline{a}_1 + n_2 \underline{a}_2 + n_3 \underline{a}_3$$

yields

$$\underline{k} \cdot \underline{t} = \frac{2\pi r_1 n_1}{N_1} + \frac{2\pi r_2 n_2}{N_2} + \frac{2\pi r_3 n_3}{N_3} = k_1 n_1 a_1 + k_2 n_2 a_2 + k_3 n_3 a_3.$$

This particular form requires

$$\underline{k}_i \cdot \underline{a}_j = k_i a_i \delta_{ij}$$

or

$$\underline{b}_i \cdot \underline{a}_j = \delta_{ij}$$

which gives

$$\underline{b}_1 \equiv \frac{\underline{a}_2 \times \underline{a}_3}{\underline{a}_1 \times \underline{a}_2 \cdot \underline{a}_3}$$

and cyclic permutations of the indices (123). With this in mind we may write for the irreducible representations of $\mathfrak{J}$

$$\Gamma^{(\underline{k})}(\underline{t}) = e^{i\underline{k} \cdot \underline{t}}$$

or more explicitly

$$\Gamma^{(\underline{k})}(\ \{E \mid \underline{t}\}).$$

Bloch's Theorem. If a function $f_{\underline{k}}$ transforms under the group $\mathfrak{J}$ according to the $\underline{k}^{\text{th}}$ irreducible representation, then it can be written in the form $f_{\underline{k}} = e^{i\underline{k} \cdot \underline{x}} u_{\underline{k}}(\underline{x})$, where $u_{\underline{k}}(\underline{x})$ has full translational symmetry.

<u>Proof</u>: Let $f_{\underline{k}}$ be such a function. Then

$$\hat{P}_{\{E \mid \underline{t}\}} f_{\underline{k}}(x) = f_{\underline{k}}(\underline{x}) \Gamma^{(\underline{k})}(\{E \mid \underline{t}\}) = f_{\underline{k}}(x) e^{i\underline{k}\cdot\underline{t}}.$$

But also

$$\hat{P}_{\{E \mid \underline{t}\}} f_{\underline{k}}(\underline{x}) = f_{\underline{k}}(\underline{x} + \underline{t}).$$

Now we may write $f_{\underline{k}}$ in any case as $e^{i\underline{k}\cdot\underline{x}} u_{\underline{k}}(\underline{x})$. So we conclude that

$$f_{\underline{k}}(\underline{x}+\underline{t}) = e^{i\underline{k}\cdot\underline{x}} e^{i\underline{k}\cdot\underline{t}} u_{\underline{k}}(\underline{x}+\underline{t}) = f_{\underline{k}}(x) e^{i\underline{k}\cdot\underline{t}} = e^{i\underline{k}\cdot\underline{x}} u_{\underline{k}}(x) e^{i\underline{k}\cdot\underline{t}}$$

Hence, $u_{\underline{k}}(\underline{x} + \underline{t}) = u_{\underline{k}}(\underline{x})$, Q.E.D.

Such a function is called a "Bloch function," and when it has been written explicitly as $e^{i\underline{k}\cdot\underline{x}} u_{\underline{k}}(\underline{x})$, it is said to be in "Bloch form."

THE BRILLOUIN ZONE

The k-vector, like the translation vector $\underline{t}$, is permitted to take on only discrete values. We recall that in our definition of k_1, $k_1 = \dfrac{2\pi r_1}{L_1}$, r_1 was one of a set of $N_1 = \dfrac{L_1}{a_1}$ integers labeling the N_1 distinct $N_1\underline{\text{th}}$ roots of unity $e^{2\pi i \frac{r}{N_1}}$. Now evidently the choice of the set $\{r_1\}$ is very arbitrary, and therefore the choice of set $\{k_1\}$ is very arbitrary. We may discard any particular r_1 from the set provided that we replace it by $r_1 + r_1' N_1$, since $e^{\frac{2\pi i}{N_1}(r_1 + r_1' N_1)} = e^{2\pi i \frac{r_1}{N_1}}$ for integral r_1'. It is convenient, although by no means necessary, to choose the set of $\{r_1\}$ to include <u>all</u> the integers between specified limits. Thus, for example, we might choose all integers from one through N_1, or from zero to $N_1 - 1$, or from $-\dfrac{N_1}{2}$ to

$\frac{N_1}{2}$-1, if N_1 is even, and so forth. These ranges on r_1 produce the corresponding ranges on k_1: $\frac{2\pi}{L_1} \leq k_1 \leq \frac{2\pi N_1}{L_1} = \frac{2\pi}{a_1}$; $0 \leq k_1 \leq \frac{2\pi}{a_1} - \frac{2\pi}{L_1}$; $- \frac{\pi}{a_1} \leq k_1 \leq \frac{\pi}{a_1} - \frac{2\pi}{L_1}$. In the same fashion, the ability to replace any r_1 by $r_1 + r_1' N_1$ is equivalent to replacing any k_1 by $k_1 + r_1' \frac{2\pi}{a_1}$, since $e^{ik_1 a_1} = e^{i(k_1 + r_1' \frac{2\pi}{a_1})a_1}$.

Now let us turn to translations in a three-dimensional lattice. Since $\mathfrak{J}$ is a direct product of three simple translation groups with N_1, N_2, and N_3 different elements, and therefore has as many different irreducible representations, we know that there must be $N_1 \times N_2 \times N_3$ different irreducible representations of $\mathfrak{J}$. In other words the k-vector may take on only $N_1 \times N_2 \times N_3$ different values. Furthermore, we know that the spacing of the component of $\underline{k}$ along $\underline{b}_1$ must be $\frac{2\pi}{L_1}$; along $\underline{b}_2$, $\frac{2\pi}{L_2}$; and along $\underline{b}_3$, $\frac{2\pi}{L_3}$. The only thing that remains therefore, is to choose a convenient boundary to limit the range of $\underline{k}$.

To complete this task we note that the $\underline{k}^{\text{th}}$ irreducible representation of $\{E \mid \underline{t}\}$ is $e^{i\underline{k} \cdot \underline{t}}$; so $\underline{k} + \underline{K}$ labels the exact same representation if $\underline{K} \cdot \underline{t} = 2\pi$. Let us write $\underline{K} = K_1\underline{b}_1 + K_2\underline{b}_2 + K_3\underline{b}_3$. Then $2\pi = K_1 n_1 + K_2 n_2 + K_3 n_3$ for arbitrary $\underline{t}$ which is to say arbitrary n_1, n_2, and n_3. Hence if $\underline{K} = r_1' 2\pi\underline{b}_1 + r_2' 2\pi\underline{b}_2 + r_3' 2\pi\underline{b}_3$, where the r_i' are any integers, then the vector $\underline{k} + \underline{K}$ labels no new representation of $\mathfrak{J}$. The set of vectors $\underline{g}_i = 2\pi\underline{b}_i$ are called "primitive reciprocal lattice vectors," and any vector $\underline{K}$ of the type $r_1'\underline{g}_1 + r_2'\underline{g}_2 + r_3'\underline{g}_3$ is called a "reciprocal lattice vector." The essential property of $\underline{K}$ is that $e^{i\underline{K} \cdot \underline{t}} = 1$ for any $\underline{t}$.

Evidently we must choose a convenient <u>primitive cell</u> within the reciprocal lattice in order to establish limits on the range of k-vectors so that all are included and there is no duplication. Then all equivalent k-vectors may be reached by a reciprocal lattice translation. The Brillouin Zone (B.Z.) is such a cell, and it is defined in the reciprocal lattice in exactly the same fashion as was the Wigner-Seitz cell in the crystal lattice.

Several points are worth underscoring here. (1) Once
the Brillouin Zone, or some other primitive cell, is agreed
upon we may, so to speak, annihilate the rest of the reciprocal lattice and leave $\underline{k}$ completely undefined except in
that cell. We can do this because the k-vector does nothing
more than label the irreducible representations of $\mathfrak{J}$,
and that is accomplished within one primitive cell. The
k-vector, from the point of view of group theory, does not
have any meaning apart from this. On the other hand, if
we $\underline{want}$ to keep the rest of the reciprocal lattice as a
set of "extra sheets" of the Brillouin Zone, we are free
to do so, of course. Suppose we have two distinct sets of
$N_1 \times N_2 \times N_3$ Bloch functions transforming according to all
of the irreducible representations of $\mathfrak{J}$. It may be convenient then to label the first set with k-vectors in the
(First) Brillouin Zone and the second set with equivalent
k-vectors in a different primitive cell, perhaps the Second
Brillouin Zone, defined as the primitive cell which sur-
rounds the First Brillouin Zone and is circumscribed by
planes as before. But we should always remember that the
function labeled by $\underline{k} + \underline{K}$ transforms according to the same
irreducible representation as the one labeled by $\underline{k}$.
(2) $Wi\underline{th}in$ the Brillouin Zone lies another lattice, namely,
the end-points of each of the k-vectors. The primitive
cell of this lattice is of characteristic linear dimension
$\frac{2\pi}{L}$ and that of the Brillouin Zone is $\frac{2\pi}{a}$ --very much larger.
Since we always think of the limit L $\rightarrow \infty$ we also think
of the k-vectors as distributed virtually continuously in
the Brillouin Zone. And since we can choose the symmetry
of this very fine-celled lattice to be no different from
that of the Brillouin Zone, we quickly lose interest in it.

   Comment: Suppose that the entire group of the
Schrödinger equation of a system were the translation
group $\mathfrak{J}$. According to the theorems and the discussion of
pages 48-57. This implies that the eigenfunctions of the Ham
iltonian will transform according to the irreducible representations of $\mathfrak{J}$, except perhaps in accidental degeneracy;

and even there new linear combinations may be formed which
transform according to irreducible representations of $\mathfrak{I}$.
In other words, the eigenfunctions of the Hamiltonian
generate irreducible representations of $\mathfrak{I}$ for normal degeneracy or else representations which may be reduced to
these for accidental degeneracy. In short, the eigenfunctions are either Bloch functions automatically or else
they may be put into Bloch form. The normally degenerate
levels must be, in fact, non-degenerate — apart from spin
degeneracy — since all the irreducible representations
are one-dimensional. And so each energy level itself may
be labeled with the same k-vector which labels the irreducible representation of $\mathfrak{I}$ according to which its eigen-
function transforms. Thus $\hat{H}\psi_{\underline{k}}(\underline{x}) = E_{\underline{k}}\psi_{\underline{k}}(\underline{x})$. In the usual
limit as $L \to \infty$, $E_{\underline{k}} \to E(\underline{k})$ where $\underline{k}$ is a continuous variable.
The equation $E(\underline{k}) = $ constant defines a two-dimensional
surface of constant energy in reciprocal space or, more
simply, **k**-space. Much of solid state physics is devoted
to determining these surfaces, their topologies, and their
spacings. They and their eigenfunctions play the same
role there that the Hartree energy levels and eigenfunctions play in atomic physics.

SPACE GROUPS WITH POINT-GROUP OPERATIONS

The previous discussion includes the entire symmetry of a crystal only if it is a triclinic system without even an inversion center. If it has any higher symmetry, the space group will contain operations of the type
$\{R \mid \underline{\tau}\}$. Our purpose now is to include this symmetry in
the problem from the start and to learn what effect it
has on our previous conclusions. By construction, the
translation group $\mathfrak{I}$ is a subgroup of $\mathfrak{G}$. We have proved
on page **142** that in fact $\mathfrak{I}$ is an invariant subgroup, and
the proof was general —valid for any of the 230 space
groups.

Now since $\mathfrak{I}$ is a subgroup of $\mathfrak{G}$ we may always choose the irreducible representations of $\mathfrak{G}$ to be such that the representations of the elements $\{E \mid \underline{t}\}$ are already in block form; for if we removed all the rotational symmetry from $\mathfrak{G}$, then we would surely be able to reduce any representation of the translation elements into block form. Let us therefore begin with that form and proceed to investigate the representation of a general element $\{R \mid \underline{\tau}\}$.

To put it another way, in generating the irreducible representations of $\mathfrak{G}$, we can always restrict the basis functions to Bloch functions. Perhaps it should be pointed out that we have done this type of thing before. When we wrote the basis functions transforming according to D^2, the d-functions, as xy, yz, zx, and $3z^2-r^2$, x^2+y^2, we were selecting those functions which would generate representations of the cubic group O_h, a subgroup of the full rotational group, which are already in block form and so reducible by inspection. This choice is <u>always possible</u>. The translational symmetry itself <u>does not require</u> the generating functions to be Bloch functions. Linear combinations of them would do as well whenever more than one function is required to generate the representation. Bloch functions are required only when $\mathfrak{G} = \mathfrak{I}$, but they are a very convenient choice in any case.

IRREDUCIBLE REPRESENTATIONS OF $\mathfrak{G}$

Rather than discuss these in full generality, we shall show how to construct them beginning from some rather arbitrary basis set of functions which we now require to be in Bloch form. For the present we shall limit ourselves to <u>symmorphic space groups</u>, in which there always exists an origin of coordinates such that every element of $\mathfrak{G}$ may be written $\{R \mid \underline{t}\} = \{E \mid \underline{t}\}\{R \mid 0\}$ and $\underline{t}$ is a lattice translation. Later we shall see how to generalize our results. We begin with a particular Bloch function $\psi_{\underline{k}}$, and

we operate upon it with the point group $\mathcal{R}$ of elements
$\{R \mid 0\}$ in order to generate new functions. These must,
as we have seen from general considerations and shall soon
prove again, also be Bloch functions.

$$\hat{P}_{\{R \mid 0\}} \psi_{\underline{k}}(\underline{x}) = P_{\{R \mid 0\}} e^{i\underline{k}\cdot\underline{x}} u_{\underline{k}}(\underline{x})$$

$$= e^{i\underline{k}\cdot R^{-1}\underline{x}} u_{\underline{k}}(R^{-1}\underline{x})$$

Since R is an orthogonal transformation for which $R\underline{x}\cdot R\underline{y} = \underline{x}\cdot\underline{y}$, we may write $\underline{k}\cdot R^{-1}\underline{x} = R\underline{k}\cdot\underline{x}$. Hence we may write

$$\hat{P}_{\{R \mid 0\}} \psi_{\underline{k}}(\underline{x}) = e^{iR\underline{k}\cdot\underline{x}} u_{\underline{k}}(R^{-1}\underline{x}) = \Phi(\underline{x}).$$

To prove that $\Phi(\underline{x})$ is a Bloch function, we operate with
$\hat{P}_{\{E \mid \underline{t}\}}$.

$$\hat{P}_{\{E \mid \underline{t}\}} \Phi(\underline{x}) = e^{iR\underline{k}\cdot(\underline{x} + \underline{t})} u_{\underline{k}}(R^{-1}\underline{x}+R^{-1}\underline{t})$$

Now by definition $u_{\underline{k}}(\underline{x} + \underline{t}') = u_{\underline{k}}(\underline{x})$; hence it follows
at once that $u_{\underline{k}}(R^{-1}\underline{x} + \underline{t}') = u_{\underline{k}}(R^{-1}\underline{x})$. This is true for
any $\underline{t}'$ and so it is true for $R^{-1}\underline{t}$ which is a lattice trans-
lation by definition of R. Hence $u_{\underline{k}}(R^{-1}\underline{x} + R^{-1}\underline{t}) = u_{\underline{k}}(R^{-1}\underline{x})$
$\equiv u'_{R\underline{k}}(\underline{x})$. Thus with this last definition we have the result
that $u'_{R\underline{k}}(\underline{x}) = u'_{R\underline{k}}(\underline{x} + \underline{t})$. Therefore

$$\hat{P}_{\{E \mid \underline{t}\}} \Phi(\underline{x}) = e^{iR\underline{k}\cdot\underline{t}} \Phi(\underline{x}),$$

and Φ is a Bloch function. Q.E.D.

We shall label all the distinct Bloch functions generated
in this manner $\psi_{R\underline{k}}$, since they generate the $R\underline{k}^{th}$ irreducible
representation of $\mathfrak{I}$. The set of distinct vectors in k-
space $R\underline{k}$, generated from one k-vector and all possible R,
is called the "star of $\underline{k}$."

Let us suppose for the simplest case, that we generate exactly h distinct k-vectors in the star of a certain k-vector and therefore h distinct Bloch functions ψ_{Rk}, where h is the order of the <u>point</u> group $\mathcal{R}$. Any k-vector with such a star is called a "general" k-vector, and its end point is called a "general point in the Brillouin Zone." We prove now that the set $\{\psi_{Rk}\}$ spans a subspace invariant under all the operations of $\overline{\mathcal{G}}$. Let the general operation be $\{S \mid \underline{t}\}$.

$$\hat{P}_{\{S \mid t\}}\psi_{R\underline{k}} = \hat{P}_{\{S \mid \underline{t}\}}\hat{P}_{\{R \mid 0\}}\psi_{\underline{k}} \qquad \text{(definition of } \psi_{R\underline{k}})$$

$$= \hat{P}_{\{E \mid \underline{t}\}}\hat{P}_{\{S \mid 0\}}\hat{P}_{\{R \mid 0\}}\psi_{\underline{k}} \qquad \text{(symmorphic group)}$$

$$= \hat{P}_{\{E \mid \underline{t}\}}\hat{P}_{\{SR \mid 0\}}\psi_{\underline{k}} \qquad \text{(law of combination)}$$

$$= \hat{P}_{\{E \mid \underline{t}\}}\psi_{SR\underline{k}} \qquad \text{(definition of } \psi_{SR\underline{k}})$$

$$= e^{iSR\underline{k}\cdot\underline{t}}\,\psi_{SR\underline{k}} \qquad (\psi_{SR\underline{k}} \text{ is a Bloch function)}$$

Since $\psi_{SR\underline{k}}$ is in the basis set (since $\{S \mid 0\}\{R \mid 0\}=\{SR \mid 0\}$ is in $\mathcal{R}$), we see that for all $\{S \mid t\}$ the subspace is invariant. It is also irreducible, because if it were not so, reduction to irreducible form would require forming linear combinations of distinct Bloch functions, which would therefore not be Bloch functions themselves, and we have already seen that the presence of the translation subgroup makes it possible to rule out all basis functions which are not Bloch functions. Consequently, the set of h distinct Bloch functions $\{\psi_{Rk}\}$ for a general $\underline{k}$ and all R in $\mathcal{R}$ generates an h-dimensional irreducible representation of $\mathcal{G}$. The representation is completely specified by the label $\underline{k}$, but it must be pointed out that it could equally well be labeled by any other $\underline{k}'$ in the star of $\underline{k}$, since the entire set of basis functions may be generated from any one.

Before we look at special points in the Brillouin Zone, let us consider the consequences of this result for a general k-vector. First we note that if we put S = E, we generate the representation of $\{E \mid \underline{t}\}$, a lattice translation, which is diagonal, as promised initially. Second, if we keep S general and put $\underline{t} = 0$, we generate the representation of $\{S \mid 0\}$, an element of $\Re$, and we see that every element of every row (or column) of the representation matrix is zero except for one element which is equal to unity. This representation is very much like the regular representation (see page 42). Third, the matrix of $\{S \mid \underline{t}\}$ is merely the matrix product of the diagonal matrix of $\{E \mid \underline{t}\}$ with that of $\{S \mid 0\}$. It therefore has the same pattern of zeros as that of $\{S \mid 0\}$. It is easy to see that $\chi(\{S \mid \underline{t}\}) = 0$ unless S = E.

If the symmetry group of the Hamiltonian of a system is the full space group $\mathcal{G}$, then the eigenfunctions may all be taken to be Bloch functions. But an eigenfunction labeled by a general k-vector will now necessarily be degenerate with h-1 other eigenfunctions (and perhaps more in the case of accidental degeneracy.) But there is a convention universally used in solid state physics which does not use the term "degeneracy" to describe this situation, and the motivating spirit behind it is to preserve the Brillouin Zone. The Brillouin Zone, we have seen, is perfectly appropriate for tabulating the irreducible representations of $\mathcal{J}$, and we have seen that the surfaces, $E(\underline{k})$ = constant, were k-space surfaces in the Brillouin Zone. On the other hand, when point group symmetries are added, we find the $\psi_{\underline{k}}$ and $\psi_{R\underline{k}}$ are degenerate - that $E(\underline{k}) = E(R\underline{k})$ for all R in $\Re$. There are two alternatives here. One is simply to reduce the Brillouin Zone to a portion of the original $\frac{1}{h}$ its volume such that we can generate the original zone from all the k-vectors in the small zone by operating upon them with all the point-group operations in $\Re$. Then h distinct Bloch functions are associated with each k-vector in this reduced zone, and $E(\underline{k})$ is only defined there and is h-fold degenerate.

The other alternative, and the one always taken in solid state physics, is to keep the original Brillouin Zone and to associate a single Bloch function with each k-vector. $E(\underline{k})$ is then defined over the entire zone, but now the extra point-group symmetries are reflected in that $E(\underline{k}) = E(R\underline{k})$ for all R in $\mathfrak{R}$. In other words, $E(\underline{k})$ has the full symmetry of $\mathfrak{R}$ if the operations are applied in k-space. But it is no longer considered to be degenerate unless there are two eigenfunctions with the same energy and the exact same k-vector. We have seen that if spin is neglected, this can never occur at general points in the Brillouin Zone if we have normal degeneracy, but we shall soon see that it may occur at special points. A degeneracy at a special point which is made necessary by symmetry is called an "essential" degeneracy. A degeneracy at any point not required by symmetry is called "accidental," as usual.

Special Points of the Brillouin Zone. Thus far we have generated all irreducible representations of $\mathfrak{G}$ which can be labeled by a general k-vector. Although the set of special k-vectors is a set of measure zero compared with the whole set, the case of these special points is of greater interest and practical importance than all the rest.

A point in the Brillouin Zone is called special if there is an operation $R \neq E$ of the point group such that $e^{iR\underline{k}\cdot\underline{t}} = e^{i\underline{k}\cdot\underline{t}}$ — in other words, if $R\underline{k} = \underline{k} + \underline{K}$, where $\underline{K}$ is any reciprocal lattice vector, including zero. If $\underline{K} = 0$, then $\underline{k}$ lies on a rotation axis or a mirror plane or both. If $\underline{K} \neq 0$, then $\underline{k}$ lies on a Brillouin Zone boundary.

Consider now the set of all elements of $\mathfrak{G}$ which, acting on a Bloch function of index $\underline{k}$, leave the index invariant; that is they generate a perhaps different Bloch function of the same index $\underline{k}$:

$$\hat{P}_{\{R \mid \underline{t}\}} \psi_{\underline{k}} = \psi_{R\underline{k}} = \psi_{\underline{k}}'.$$

Unless $\underline{k}$ is a special point, the only operations with this property are the group of translations. But at the special points, where $R\underline{k} = \underline{k} + \underline{K}$, the Bloch function index $\underline{k} + \underline{K}$ is the same as the index $\underline{k}$. We now prove that the set of all such operations forms a group. The identity element is $\hat{P}_{\{E \mid 0\}}$. $\hat{P}_{\{R \mid \underline{t}\}}^{-1} \hat{P}_{\{R \mid \underline{t}\}} \psi_{\underline{k}} = \hat{P}_{\{R \mid \underline{t}\}}^{-1} \psi_{\underline{k}}'$ and so $\hat{P}_{\{R \mid \underline{t}\}}^{-1} \psi_{\underline{k}}' = \psi_{\underline{k}}$, which means that $\hat{P}_{\{R \mid \underline{t}\}}^{-1}$ is one of these operations if $\hat{P}_{\{R \mid \underline{t}\}}$ is. Similarly $\hat{P}_{\{R' \mid \underline{t}'\}} \hat{P}_{\{R \mid \underline{t}\}}$ is in the group if $\hat{P}_{\{R' \mid \underline{t}'\}}$ and $\hat{P}_{\{R \mid \underline{t}\}}$ are.

The set of all elements of $\mathcal{G}$ with this property is called the "small group of $\underline{k}$," or simply the "group of $\underline{k}$," and we shall symbolize it by $\mathcal{K}$. (Note that $\mathcal{K}$ is always defined with respect to a particular k-vector.) The translation subgroup $\mathcal{J}$ is always contained in $\mathcal{K}$ and is, as we showed, an invariant subgroup of $\mathcal{G}$ and therefore of $\mathcal{K}$ as well. Consequently we may form the factor group of $\mathcal{K}$ taken with respect to $\mathcal{J}$ and, as we have proved on page 143, it is isomorphic with the point group $\mathcal{R}_{\underline{k}}$ consisting of all operations of the form $\{R \mid 0\}$ which are in $\mathcal{K}$. (This last statement is only true of the symmorphic space groups. In the non-symmorphic groups, $\mathcal{K}/\mathcal{J}$ is still isomorphic with $\mathcal{R}_{\underline{k}}$, but the later is not part of $\mathcal{K}$.) Let us attempt, therefore, to build up the irreducible representations of $\mathcal{G}$ out of the irreducible representations of $\mathcal{J}$ and of $\mathcal{R}_{\underline{k}}$.

Let us begin with an arbitrary Bloch function of index $\underline{k}$ and operate on it with all of the elements of $\mathcal{R}_{\underline{k}}$. Since $\mathcal{R}_{\underline{k}}$ is contained in $\mathcal{K}$, the index will be invariant and the set of distinct Bloch functions thus generated will transform according to some representation of $\mathcal{R}_{\underline{k}}$. We may then reduce this representation by forming new linear combinations among these Bloch functions; but since they all have the same index $\underline{k}$, any linear combination among them will also be a Bloch function of index $\underline{k}$. Let us therefore choose from the start a specific set of Bloch functions which already transform according to an irreducible representation of $\mathfrak{R}_{\underline{k}}$. We label one of these functions $\psi_{\underline{k}\lambda}^{(i)}(\underline{x})$, where the $\overline{i}$ labels the irreducible representation of $\mathfrak{R}_{\underline{k}}$; $\underline{k}$ labels the irreducible representation of $\mathfrak{I}$; and λ labels the particular function which transforms under $\mathfrak{R}_{\underline{k}}$ according to the $\lambda^{\underline{th}}$ column of the representation matrix:

$$\hat{P}_{\{R_{\underline{k}} \mid 0\}} \psi_{\underline{k}\lambda}^{(i)} = \psi_{\underline{k}\mu}^{(i)} \Gamma^{(i)}(\{R_{\underline{k}} \mid 0\})_{\mu\lambda}$$

If we operate with all operations $\{R \mid 0\}$ of $\mathfrak{R}$ on these functions $\{\psi_{\underline{k}\lambda}^{(i)}(\underline{x})\}$, or more specifically with the operations in the cosets

$$\{E \mid 0\} \; \mathfrak{R}_{\underline{k}} \; , \; \{R_1 \mid 0\}\mathfrak{R}_{\underline{k}}, \; \ldots, \; \{R_n \mid 0\}\mathfrak{R}_{\underline{k}},$$

we generate n different sets, covering the star of the k-vector, which span an irreducible subspace invariant under all operations of $\mathfrak{G}$. The total number of sets n is equal to $h/h_{\underline{k}}$, where h is the order of the point group $\mathfrak{R}$ and $h_{\underline{k}}$ is the order of the group of $\underline{k}$ $\mathfrak{R}_{\underline{k}}$. These sets form the basis for irreducible representations of the full space group $\mathfrak{G}$.

In practice we shall have little use for the full representation matrices of $\mathfrak{G}$. They are completely specified by a k-vector and by a particular irreducible representation of the group $\mathfrak{R}_{\underline{k}}$. Thus usually we need know nothing more than the representations and character tables of $\mathfrak{R}_{\underline{k}}$ for every $\underline{k}$ in the Brillouin Zone. For a general k-vector $\mathfrak{R}_{\underline{k}} = \{E \mid 0\}$, and so the representation is completely labeled by $\underline{k}$.

Degeneracy at Special Points of the Brillouin Zone.
We have seen that the dimension of an irreducible representation of $\mathfrak{G}$ specified by $\underline{k}$ and the $i^{\underline{th}}$ irreducible repre-

sentation of $\mathcal{R}_{\underline{k}}$ is $\frac{h}{h_{\underline{k}}} \times \ell_i$, where h is the order of $\mathcal{R}$, $h_{\underline{k}}$ is the order of $\mathcal{R}_{\underline{k}}$, and ℓ_i is the dimension of the $i^{\underline{th}}$ irreducible representation of $\mathcal{R}_{\underline{k}}$. But $\frac{h}{h_{\underline{k}}}$ is also equal to the number of k-vectors in the star of $\underline{k}$. Now we recall from the discussion on page 154 that the energy eigenvalue was labeled by $\underline{k}$, and $\underline{k}$ was permitted to range over the entire Brillouin Zone, so that even though $E(\underline{k}) = E(R\underline{k})$, $\psi_{\underline{k}}$ and $\psi_{R\underline{k}}$ were not called degenerate eigenfunctions. This same convention applies to the special points, but now it is possible, when $\ell_i \geq 2$, to have degenerate eigenfunctions with the same k-vector. This is called an "essential" degeneracy, as we said earlier, and its degree is equal to ℓ_i. We shall see later that such degeneracy means that energy bands $E_i(\underline{k})$, i = 1,2,...,ℓ_i, which are necessarily separate from one another at general points in k-space may all come together into an ℓ_i-fold degenerate level at a special point.

EXAMPLE I: A SYMMORPHIC GROUP

In order to give concrete expression to these ideas, let us consider an example. To keep it very simple but to display most of the properties of more complicated systems, we consider a 2-dimensional square lattice of point ions.

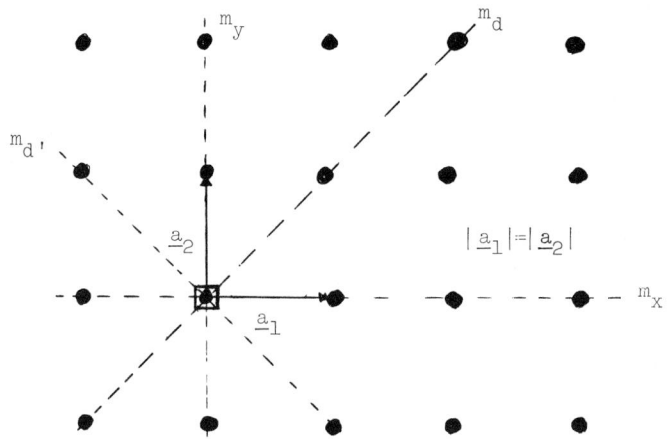

Let us begin to enumerate the space-group elements which
leave this system invariant. With the usual periodic
boundary conditions, any lattice translation $\{E \mid \underline{t}\}$, where
$\underline{t} = n_1\underline{a}_1 + n_2\underline{a}_2$, is such an operation. To enumerate the
point-group operations, we pick an origin at a lattice
site, and we find the following symmetry operations:
$\{E \mid 0\}$, $\{C_4 \mid 0\}$, $\{C_4^{-1} \mid 0\}$, $\{C_4^2 \mid 0\}$, $\{m_x \mid 0\}$, $\{m_y \mid 0\}$,
$\{m_d \mid 0\}$, $\{m_{d'} \mid 0\}$. (In order to keep the problem simple
let us suppose that the ions have a vector-like property
in the z-direction, and so rule out a mirror plane and all
2-fold axes contained in the xy-plane.) Now let us have a
careful look at other symmetry points in the lattice to
ascertain whether the symmetry operations which may be
performed there are of the form $\{R \mid \underline{t}\}$ or not, where R is
one of the point-group operations already found. First of
all, we note that there is a mirror plane parallel to m_y
but displaced $\frac{1}{2} \underline{a}_1$ from it. We might anticipate that the
space-group operation would therefore involve half a lat-
tice translation and so be an entirely new element in the
space group; but we would be wrong. For we must express
this operation with respect to <u>the same origin</u> from which
the original point-group operations were defined. Now
the rule for performing a specific operation $\{R \mid \underline{t}\}$ about
an origin, displaced an amount τ away from the old one, is
given by the usual kind of transformation : $\{E \mid \tau\} \{R \mid \underline{t}\}$
$\{E \mid \tau\}^{-1}$ is the operation. In our case the old operation
is $\{m_y \mid 0\}$, and it is to be performed with respect to an
origin displaced a distance $\frac{\underline{a}_1}{2}$ from the old. So the new
operation is

$$\{E \mid \tfrac{\underline{a}_1}{2}\}\{m_y \mid 0\}\{E \mid -\tfrac{\underline{a}_1}{2}\} = \{E \mid \tfrac{1}{2}\underline{a}_1\}\{m_y \mid +\tfrac{1}{2}\underline{a}_1\} = \{m_y \mid \underline{a}_1\}$$

but this <u>is</u> in the form $\{R \mid \underline{t}\}$. It is easy to confirm
that the operation $\{m_y \mid 0\}$, with respect to the old origin
and the translation $\{E \mid \underline{a}_1\}$, is exactly the same operation
as the reflection performed about a point $\frac{1}{2}\underline{a}_1$ to the right

of the origin.

In the same way we can see that every operation of the point group may also be performed about the point $\frac{1}{2}\underline{a}_1 + \frac{1}{2}\underline{a}_2$ from the origin. But by writing $\{E \mid \frac{1}{2}\underline{a}_1 + \frac{1}{2}\underline{a}_2\}$ $\{R \mid 0\}\{E \mid -\frac{1}{2}\underline{a}_1 -\frac{1}{2}\underline{a}_2\}$ we can readily see that the new operations are all of the form $\{R \mid \underline{t}\}$, where $\underline{t} = 0$, or $\underline{a}_1$, or $\underline{a}_2$. Finally, there is one operation which has every appearance of being totally new. It is a reflection in a mirror parallel to m_d but displaced an amount $\frac{1}{4}(\underline{a}_1 -\underline{a}_2)$ from it, followed by a translation by an amount $\frac{1}{2}(\underline{a}_1 + \underline{a}_2)$. Such an operation is a <u>glide plane</u>. But let us apply the test: new operation = $\{E \mid \frac{1}{4}\underline{a}_1 -\frac{1}{4}\underline{a}_2\}\{m_d \mid \frac{1}{2}\underline{a}_1 +\frac{1}{2}\underline{a}_2\}$ $\{E \mid -\frac{1}{4}\underline{a}_1 +\frac{1}{4}\underline{a}_2\}$ = $\{E \mid \frac{1}{4}\underline{a}_1 -\frac{1}{4}\underline{a}_2\}\{m_d \mid -\frac{1}{4}m_d\underline{a}_1 +\frac{1}{4}m_d\underline{a}_2 +\frac{1}{2}\underline{a}_1 +\frac{1}{2}\underline{a}_2\}$ = $\{E \mid \frac{1}{4}\underline{a}_1 -\frac{1}{4}\underline{a}_2\}\{m_d \mid -\frac{1}{4}\underline{a}_2 +\frac{1}{4}\underline{a}_1 +\frac{1}{2}\underline{a}_1 +\frac{1}{2}\underline{a}_2\}$ = $\{m_d \mid \underline{a}_1\}$. And so this "new" operation is already included in the set $\{R \mid \underline{t}\}$.

In this way we conclude that the space group is symmorphic, and its point group $\Re$ is the one we have called 4mm or C_{4v}. Now let us construct the Brillouin Zone. The shortest reciprocal lattice vectors are

$$2\pi\underline{b}_1 = \frac{2\pi\frac{\underline{a}}{2}1}{a_1^2} = \underline{K}_1 \text{ and } 2\pi\underline{b}_2 = \frac{2\pi\underline{a}}{a_2^2}2 = \underline{K}_2$$

We take $\underline{K}_1$ and $\underline{K}_2$ as primitive vectors and, by choosing an arbitrary site, we construct all the reciprocal lattice vectors and all their perpendicularly bisecting planes. The smallest enclosed region is a square, and it is the (First) Brillouin Zone.

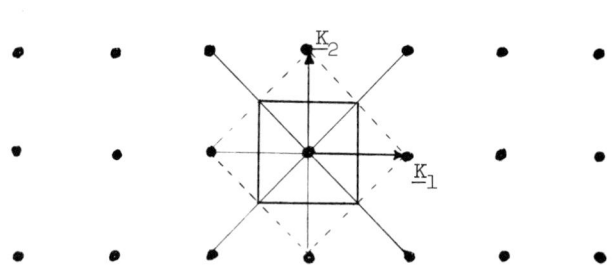

The next enclosed region contains k-vectors, all of which
are equivalent to those in the first region as may be seen
by the fact that each of the triangular portions of the
second region may be translated into a corresponding portion of the first by means of a translation from the set
$\pm K_1$, $\pm K_2$. This second region is called the Second
Brillouin Zone, page 149. Let us reproduce now the First
Brillouin Zone to find the special points of the zone.

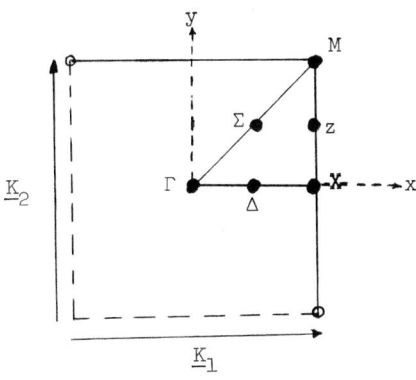

First we must establish some convention for dealing with
the edges (and corners) of the Brillouin Zone, since the
k-vectors which lie there differ from others which also
lie on the edges (and corners) by a reciprocal lattice
vector, and so must be in the same Bloch state. Our
convention is to keep in the Brillouin Zone only those which
lie on the solid lines and only the one at M. There are
six classes of special points in this zone. The one at Γ
and the one at M have the same group of $\underline{k}$, equal in this
case to the entire point group C_{4v}. The star of $\underline{k}$ in
each case has but one k-vector. The point at X is left
invariant by E, m_x, m_y, and C_4^2. (Note that the point at X
is $\underline{k} = \frac{1}{2}\underline{K}_1$, so that $m_y\underline{k} = -\underline{k} = -\frac{1}{2}\underline{K}_1 = \frac{1}{2}\underline{K}_1 - \underline{K}_1 = \underline{k} - \underline{K}_1$ and so
is equivalent to $\underline{k}$.) The star of $\underline{k}$ in this case has but
two points; the group of $\underline{k}$ is 2mm or C_{2v}.

The label Δ stands for any point, except the end
points, of the line between Γ and X. Its star evidently

has four points; so the group of $\underline{k}$ (or the group of Δ, as it may also be called) has but two elements, E and m_x. Similarly the group of Σ is E and m_d, and that of Z is E, m_y.

The star of the general k-vector is shown in the figure, and since it has 8 points the group of $\underline{k}$ can contain only E.

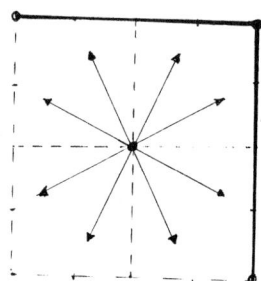

We might ask perhaps why the midpoint of the upper boundary is not taken to be a special point, since it is not essentially different from X, and the answer is that it might indeed have been chosen. But then X would have been omitted. The reason is that the set of Bloch functions which generate the irreducible representation of $\mathfrak{G}$, labeled by $\underline{k}_X$ and an irreducible representation of X, contain functions of index $\underline{k}_X$ and also those of index $R\underline{k}_X$, where R is not in the group of X. In our case, $R\underline{k}_X$ is in fact merely the k-vector whose end point is the midpoint of the upper boundary. Hence, if we constructed all irreducible representations labeled by $\underline{k}_X$ and all labeled by $R\underline{k}_X \neq \underline{k}_X$, the two sets would be equivalent to one another.

Our guiding principle in selecting the special points has been the following: to select the smallest singly connected region of the Brillouin Zone whose interior contains only general points and from which the entire zone can be generated by applying to all its k-vectors the h operations of the point group $\mathfrak{R}$. Once we have found this portion, $\frac{1}{h}$ of the whole volume, we may forget about the rest of the zone. For example, once $E_i(\underline{k})$ is known in that sector, it is known everywhere.

In the following tables we show the characters of all small groups of $\underline{k}$ for the special points. The k-vector and the label of the irreducible representation of $\mathcal{K}$, it will be recalled, completely label the space-group representation.

Γ , M	E	$2C_4$	$C_4^{\;2}$	$m_x m_y$	$m_d m_{d'}$
Γ_1, M_1	1	1	1	1	1
Γ_2, M_2	1	1	1	-1	-1
Γ_3, M_3	1	-1	1	1	-1
Γ_4, M_4	1	-1	1	-1	1
Γ_5, M_5	2	0	-2	0	0

X	E	$C_4^{\;2}$	m_x	m_y
X_1	1	1	1	1
X_2	1	1	-1	-1
X_3	1	-1	1	-1
X_4	1	-1	-1	1

Δ	E	m_x
Δ_1	1	1
Δ_2	1	-1

Σ	E	m_d
Σ_1	1	1
Σ_2	1	-1

Z	E	m_y
Z_1	1	1
Z_2	1	-1

Simply by looking at the character tables we may conclude at once that an essential degeneracy may only occur at the points Γ and M. No energy bands come together at any other special points of the zone due to the symmetry of the space group.

Now let us examine what happens to the energy bands as we leave a point of high symmetry in the Brillouin Zone and go to one of lower symmetry. We know, for example, that a k-vector whose end point lies along the line of Δ, no matter how close it is to the point Γ or the point X, must label an irreducible representation of $\mathcal{G}$ with the additional label Δ_1 or Δ_2, depending upon how its generating functions transform under m_x. It only seems reasonable, therefore, that when $\underline{k}$ actually reaches the point Γ or the point X, the generating functions there would have the same symmetry under m_x, so that a Δ_1 band might turn into a Γ_1 or a Γ_3 band at Γ or an X_1 or X_3 band at X. Similarly Δ_2 might change its label into Γ_2, Γ_4, X_2, or X_4. Finally Δ_1 and Δ_2 may merge into a doubly degenerate Γ_5. It also seems quite plausible that all of those so-called <u>compatibility relations</u> might be derived in a manner completely analogous with the methods we used in the discussion of crystal field splittings of atomic levels (pp.93-98). Thus we would conclude that the Γ_5 level would split into a Δ_1 and a Δ_2 band along the Δ line. This is, as it happens, the right method, but our reasoning has been at best only suggestive.

To make clear the necessity of a more rigorous justification of this procedure, consider the fact that at the point Γ the space-group representation labeled Γ_5 has dimension $\frac{h}{h_k} \times \ell_i = \frac{8}{8} \times 2 = 2$, while at any Δ point the representation Δ has dimension $\frac{8}{2} \times 1 = 4$. Now in the crystal field splitting case we used the above methods when we went from fields of high symmetry to those of lower symmetry; we also went from irreducible representations of <u>higher dimension</u> to others of <u>lower dimension</u> representing the group of lower symmetry. In our case, when we leave a point of high symmetry in the Brillouin Zone to go

to one of lower symmetry, the dimension of the representations changes in the opposite sense.

The essential argument in the crystal field case
was the following: if a specific degenerate set of wave
functions generated an irreducible representation of a
Schrödinger equation group of higher symmetry, then for a
very small change in an external parameter of the potential,
these same wave functions were still nearly the exact wave
functions of the new Hamiltonian; but the change in the
external parameter, no matter how slight, destroyed the
higher symmetry. So these wave functions must generate a
generally reducible representation of the group of lower
symmetry. And so we could use the character tables to
predict the splittings. This was all done in the spirit
of perturbation theory.

But the case we are considering now is completely
different to all appearances. First we are not altering
an external parameter of the lattice potential in any way.
We do not alter the symmetry of the space group in the
slightest. The symmetry that we are speaking of is a
symmetry in $\underline{k}$-space, and the small changes we are considering are changes of a k-vector. But this is surely not an
external parameter of the system, for it is nothing more
than a label of an irreducible representation of a fixed
system. A change in $\underline{k}$ is a change between two distinct
irreducible representations of that system. The physics
of the two situations is totally different.

And yet there is a fundamental analogy between
these problems which we shall attempt to bring forth.
Recall that the fundamental building block for an irredu-
cible representation of $\mathcal{G}$ was a set of functions $\psi_{\underline{k}}^{(i)}$
which transformed according to the $i\underline{\text{th}}$ irreducible
representation of the group of $\underline{k}$, $\mathcal{K}$, and the $\underline{k}\underline{\text{th}}$ irreducible representation of $\mathcal{J}$. Thus the whole set could
be written $e^{i\underline{k}\cdot\underline{x}}u_{\underline{k}}^{(i)}(\underline{x})$, where the set $u_{\underline{k}}^{(i)}$ generates the
$i\underline{\text{th}}$ irreducible representation of $\mathcal{K}$. We saw that the
complete set of functions generating this irreducible
representation of $\mathcal{G}$ could be produced by operating on

this set with elements of $\mathfrak{R}$ not in $\mathfrak{R}_k$. Now if one of these functions is an eigenfunction of a Hamiltonian with the symmetry of $\mathfrak{G}$, then all others must be degenerate, in the older sense of the word, with it. Hence the physical problem of the periodic potential has been completely solved once the full set of $u_k^{(i)}$ and the energies $E(\underline{k})$ has been determined. Let us write the Schrödinger equation for a particular $\psi_{\underline{k}\mu}^{(i)}$:

$$H\psi_{\underline{k}\mu}^{(i)}(-\frac{\hbar^2}{2m}\nabla^2+V(\underline{x}))e^{i\underline{k}\cdot\underline{x}}u_{\underline{k}\mu}^{(i)}(\underline{x}) = E(\underline{k})\,e^{i\underline{k}\cdot\underline{x}}u_{\underline{k}\mu}^{(i)}(\underline{x})$$

$$e^{i\underline{k}\cdot\underline{x}}\{-\frac{\hbar^2}{2m}(\nabla^2+2i\underline{k}\cdot\underline{\nabla}-\underline{k}\cdot\underline{k})u_{\underline{k}\mu}^{(i)}(\underline{x})+V(\underline{x})u_{\underline{k}\mu}^{(i)}\} = E(\underline{k})e^{i\underline{k}\cdot\underline{x}}u_{\underline{k}\mu}^{(i)}(\underline{x})$$

$$(-\frac{\hbar^2}{2m}\nabla^2+V(x)+\frac{\hbar k}{m}\cdot\underline{p})u_{\underline{k}\mu}^{(i)}(\underline{x})=E'(\underline{k})u_{\underline{k}\mu}^{(i)}(\underline{x})$$

where $\underline{p} = -i\hbar\underline{\nabla}$ and $E'(\underline{k}) = E(\underline{k}) - \frac{\hbar^2k^2}{2m}$. Now since both $V(\underline{x})$ and $u_{\underline{k}\mu}^{(i)}(\underline{x})$ are translationally invariant, this last equation may be solved in a single primitive cell of the lattice with boundary conditions given by the periodicity of $u_{\underline{k}\mu}^{(i)}$ and the usual continuity conditions on any wave function and its normal component at a foundary.

The equation for $u_{\underline{k}\mu}^{(i)}$ has the appearance of an ordinary equation of the Schrödinger type but with an extra potential energy $\frac{\hbar k}{m}\cdot\underline{p}$. We may now think of some analogous physical system with such a Hamiltonian and for which $u_{\underline{k}\mu}^{(i)}(\underline{x})$ is now an eigenfunction and $E'(\underline{k})$ is the true energy eigenvalue. For such a system, $\underline{k}$ is simply an external parameter which may indeed be varied at will. Now let us investigate the group of this Schrödinger equation. We have already seen that $\mathfrak{J}$ leaves its Hamiltonian invariant. Let us consider the form which it takes when its Hamiltonian is written in terms of $\underline{x}' = R^{-1}\underline{x}$, where R is any orthogonal transformation in $\mathfrak{R}$. The Laplacian operator is, of course, invariant and so is V by definition of $\mathfrak{R}$.

The remaining term is $\frac{h}{m} \underline{k} \cdot \hat{p}(\underline{x}') = \frac{hk}{m} \cdot R^{-1}\underline{p} = \frac{h}{m} R\underline{k} \cdot \underline{p}$,
since $\underline{v}$ transforms exactly the way $\underline{x}$ does. Thus this term
will be invariant only if $R\underline{k} = \underline{k}$. <u>The group of this Schrö-</u>
<u>dinger equation, therefore, is exactly the same as $\mathcal{H}$ in</u>
<u>the real problem.</u>

Now we may certainly use the methods of crystal
field splitting for this fictitious but analogous system,
and by them derive the symmetry properties of the new
eigenfunctions when we alter $\underline{k}$ ever so slightly and destroy
a part of the original symmetry. But these new eigenfunctions must, by the precise analogy, be a new set of
$u'^{(j)}_{\underline{k}'}(\underline{x})$ from which we may build the entire irreducible
representation of $\mathcal{G}$ labeled by $\underline{k}' = \underline{k} + \delta\underline{k}$ and by the jth
irreducible representation of the group of $\underline{k}'$.

So our original surmise was correct. We may derive all compatibility relations by following the loss of
symmetry as we go from one group of $\underline{k}$ to another which is
a subgroup of the first, using exactly the methods we used
to determine crystal field splittings of degeneracy. Now
we give the results for the square lattice.

$$\Gamma_1, M_1 \rightarrow \Delta_1, \Sigma_1, Z_1 \qquad\qquad X_1 \rightarrow \Delta_1, Z_1$$

$$\Gamma_2, M_2 \rightarrow \Delta_2, \Sigma_2, Z_2 \qquad\qquad X_2 \rightarrow \Delta_2, Z_2$$

$$\Gamma_3, M_3 \rightarrow \Delta_1, \Sigma_2, Z_1 \qquad\qquad X_3 \rightarrow \Delta_1, Z_2$$

$$\Gamma_4, M_4 \rightarrow \Delta_2, \Sigma_1, Z_1 \qquad\qquad X_4 \rightarrow \Delta_2, Z_1$$

$$\Gamma_5, M_5 \rightarrow (\Delta_1 + \Delta_2), (\Sigma_1 + \Sigma_2), (Z_1 + Z_2)$$

From these results we may deduce more remote compatibili-
ties. Thus, since Γ and M are connected along Σ, we may
be sure that an energy level which has Γ_1 symmetry can
only connect at M with one of M_1, M_4, or M_5 symmetry.

EXAMPLE II: A NON-SYMMORPHIC GROUP

Let us now consider another "crystal" with the
two-dimensional lattice of point ions sketched below.

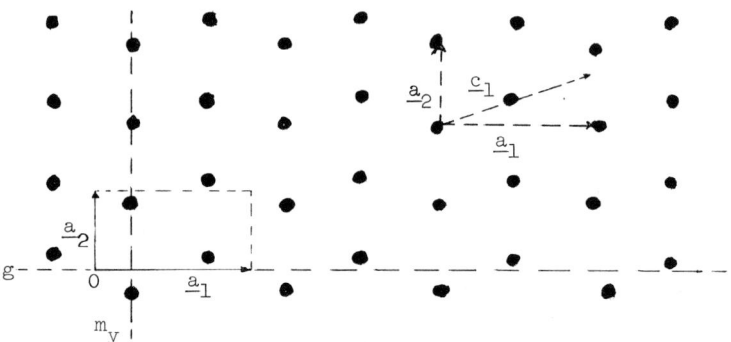

First we find a suitable set of primitive vectors. We
note that $\underline{c}_1$ is not primitive because $2\underline{c}_1$ is not equivalent
to $\underline{c}_1$, as there is no ion there. But $\underline{a}_1$ and $\underline{a}_2$ evidently
are a possible choice, and they evidently define a Bravais
lattice with rectangular symmetry since $\underline{a}_1 \cdot \underline{a}_2 = 0$. Note
however that the Bravais lattice does not coincide with
the lattice of ions; instead we must associate two ions
with each Bravais lattice point. This situation is some-
times spoken of as a Bravais lattice with a <u>basis</u>. Now we
look for the space-group operations other than the pure
translations. First we note that the point-group operations
consistent with the rectangular Bravais lattice are E, m_x,
m_y, and C_{2z}. (Once more we rule out for simplicity operations which take z into -z.) But our lattice has a basis;
so we must be careful to see how these operations appear
in the space group. We must pick an origin from which to
describe all operations, and the one we have chosen is
the one labeled "0" at the bottom of the figure. The
primitive cell appropriate to it is sketched in there and
is seen to contain two ions symmetrically placed, usually
a convenient choice. Now the point 0 is a two-fold axis

for our system; hence one space-group operation is
$\{C_{2z} \mid 0\}$. Another is the mirror m_y taken with respect to
an origin displaced $\frac{1}{4}\underline{a}_1$ or $-\frac{1}{4}\underline{a}_1$ from 0. Thus with respect
to 0, the operation is

$$\{E \mid \pm\tfrac{1}{4}\underline{a}_1\}\{m_y \mid 0\}\{E \mid \mp\tfrac{1}{4}\underline{a}_1\} = \{E \mid \pm\tfrac{1}{4}\underline{a}_1\}\{m_y \mid \pm \tfrac{1}{4}\underline{a}_1\} = \{m_y \mid \pm\tfrac{1}{2}\underline{a}_1\}.$$

Since these two operations differ by a lattice translation
we choose one of them $\{m_y \mid \tfrac{1}{2}\underline{a}_1\}$ as fundamental. Next we
note the glide plane marked g in the figure. Since the
mirror plane passes through 0, we may write at once the
operation: $\{m_x \mid \tfrac{1}{2}\underline{a}_1\}$. There are other glide planes, m_y-
mirrors, and two-fold centers, but they are all expressible
as products of these operations with some lattice translation. Thus we may take as a fundamental set, $\{E \mid 0\}$,
$\{C_{2z} \mid 0\}$, $\{m_y \mid \tfrac{1}{2}\underline{a}_1\}$, and $\{m_x \mid \tfrac{1}{2}\underline{a}_1\}$ all taken with respect
to 0.

We might think that we have here a non-symmorphic
space group, since the fundamental set is not of the form
$\{R \mid 0\}$. However we must explore the possibility that a
shift of the origin might be able to rectify this situation.
Hence let us operate on every element of our fundamental
set in order to shift the origin to some arbitrary point $\underline{\tau}$.
(Let $\underline{\tau}_1$ and $\underline{\tau}_2$ be components along $\underline{a}_1$, $\underline{a}_2$.)

$$\{E \mid -\underline{\tau}\}\{E \mid 0\}\{E \mid \underline{\tau}\} = \{E \mid 0\}$$

$$\{E \mid -\underline{\tau}\}\{C_{2z} \mid 0\}\{E \mid \tau\} = \{C_{2z} \mid -2\underline{\tau}\}$$

$$\{E \mid -\tau\}\{m_y \mid \tfrac{1}{2}\underline{a}_1\}\{E \mid \tau\} = \{m_y \mid \tfrac{1}{2}\underline{a}_1 - 2\underline{\tau}_1\}$$

$$\{E \mid -\tau\}\{m_x \mid \tfrac{1}{2}\underline{a}_1\}\{E \mid \tau\} = \{m_x \mid \tfrac{1}{2}\underline{a}_1 - 2\underline{\tau}_2\}$$

Clearly there is no $\underline{\tau}$ such that all operations taken with
respect to the new origin are of the form $\{R \mid \underline{t}\}$ where
$\underline{t} = n_1\underline{a}_1 + n_2\underline{a}_2$. Hence we <u>must</u> conclude that the group is
non-symmorphic.

Note that although some $\underline{\tau}$ may be found for each
of the first three elements such that they may separately
be put in the form $\{R \mid 0\}$ or $\{R \mid \underline{t}\}$, no $\underline{\tau}$ exists which can
do this to the fourth element. It is this property which
characterizes the glide planes of the non-symmorphic groups.

The Brillouin Zone appropriate to this Bravais
lattice is shown in the figure below together with all
the special points.

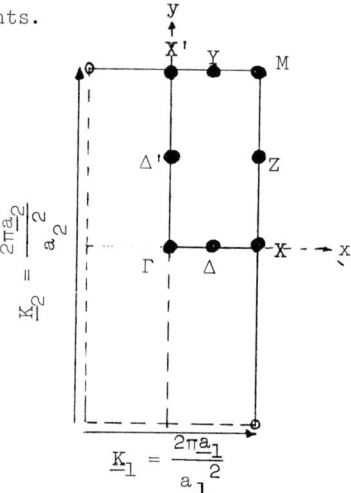

If we had this Bravais lattice without a basis we would
know how to proceed at once to analyze the qualitative
features of the band structure. But now we must return
to the fundamental problem of the irreducible representations of the non-symmorphic space groups. Our purpose
here, for the sake of expediency, will be more to sketch
proofs and display useful results than to convince with
rigor.

Let us first consider, as we did in the last example, the set of Bloch functions which may be generated
from one such function of index $\underline{k}$, where $\underline{k}$ is a general
point in the Brillouin Zone, by the successive application
of the four operations, $\{E \mid 0\}$, $\{C_2 \mid 0\}$, $\{m_y \mid \frac{1}{2}\underline{a}_1\}$, and
$\{m_x \mid \frac{1}{2}\underline{a}_1\}$. Since $\underline{k}$ is a general point we get four distinct
Bloch functions:

$$\psi_{\underline{k}_1} = e^{i\underline{k}\cdot\underline{x}}u_k(\underline{x})$$

$$\psi_{\underline{k}_2} = e^{iC_2\underline{k}\cdot x}u_k(C_2\underline{x})$$

$$\psi_{\underline{k}_3} = e^{im_y\underline{k}\cdot\underline{x}}e^{i\frac{1}{2}\underline{k}\cdot\underline{a}_1}u_{\underline{k}}(m_y\underline{x} + \tfrac{1}{2}\,\underline{a}_1)$$

$$\psi_{\underline{k}_4} = e^{im_x\underline{k}\cdot\underline{x}}e^{i\frac{1}{2}\underline{k}\cdot\underline{a}_1}u_{\underline{k}}(m_x\underline{x} + \tfrac{1}{2}\,\underline{a}_1)$$

Since every operation in $\mathbf{G}$ can be written $\{E\,|\,\underline{t}\}\{R\,|\,\underline{\tau}\}$, where $\underline{t}$ is a lattice translation and $\{R\,|\,\underline{\tau}\}$ is one of the above four elements, it is easy to confirm that these four functions span a subspace invariant under $\mathbf{G}$. Furthermore, it must be irreducible because to reduce it would require linear combinations of Bloch functions with different k-vectors, and we were permitted from the start to rule out the resulting functions because they would not be Bloch functions. Hence the set of four functions above will generate a four-dimensional representation of every element of $\mathbf{G}$, and it will be irreducible. Thus the results for the general k-vector will be no different from those of the symmorphic groups.

But suppose $\underline{k}$ is a special point. Then by definition, $\underline{k}$, $C_2\underline{k}$, $m_x\underline{k}$, and $m_y\underline{k}$ will not all be distinct; so we must again admit the possibility of reducing a representation built from these four functions when $\underline{k}$ is a special point. We might be tempted to follow the example of the symmorphic groups and begin with a set of Bloch functions, all of index $\underline{k}$, which transform according to the $i^{\underline{th}}$ irreducible representation of the point group whose elements are taken from $\{E\,|\,0\}$, $\{C_2\,|\,0\}$, $\{m_y\,|\,0\}$, and $\{m_x\,|\,0\}$, even though not all these elements are in $\mathbf{G}$ and even though this leaves $\underline{k}$ invariant. We then operate upon the set with characteristic elements which are in $\mathbf{G}$ and generate enough new functions to build a representation of $\mathbf{G}$. Fundamentally the idea is sound, and it usually works.

We may put it on a somewhat firmer basis by re-
calling that the essential simplicity of the symmorphic
groups arose from the fact that the factor group of $\mathcal{G}$,
taken with respect to $\mathfrak{J}$, was isomorphic with the point
group $\mathfrak{R}$ and, more particularly, that the factor group of
$\mathfrak{K}$, taken with respect to $\mathfrak{J}$, was isomorphic with the
point group $\mathfrak{R}_k$, and $\mathfrak{R}_k$ was a subgroup of $\mathfrak{K}$. And so it
is sufficient to take a set of Bloch functions which trans-
form according to an irreducible representation of $\mathfrak{R}_k$
and build the entire representation of $\mathcal{G}$ from them.
In the case of the non-symmorphic groups, $\mathfrak{K}/\mathfrak{J}$ is still
isomorphic with $\mathfrak{R}_k$, but the elements of $\mathfrak{R}_k$, which are of
the form $\{R \mid 0\}$, are not all now elements of either $\mathfrak{K}$ or $\mathcal{G}$.
We can still proceed and consider elements of the form
$\{R \mid \tau\}$ which belong to $\mathfrak{K}$, but these, when acting on Bloch
functions $e^{i k x} u_k(x)$, will introduce phase factors of the
form $e^{i k \tau}$, which are reminiscent of the factors (-1)
encountered in the half-integral (spin) representations of
the point groups. We can handle these phase factors by
expanding the theory of representations in order to include
a more general definition (Ray representations) such that
if

$$R \ S = T,$$

their matrix representation satisfies

$$M_R \ M_S = a_{RST} \ M_T$$

where a_{RST} is a given set of phase factors.

In our case, however, we shall solve the problem
by, once again, enlarging the factor group $\mathfrak{K}/\mathfrak{J}$ artifi-
cially and then retaining only those representations which
are compatible with Bloch functions of a specified $\underline{k}$.

We enlarge the factor group $\mathfrak{K}/\mathfrak{J}$ to a new group
$\mathfrak{K}/\mathfrak{J}_k$, where $\mathfrak{J}_k$ is an invariant subgroup of $\mathcal{G}$ (it is
also a subgroup of $\mathfrak{J}$) which includes only the translations

$\{E \mid \underline{t}_{\underline{k}}\}$ such that

$$\underline{k} \cdot \underline{t}_{\underline{k}} = 2\pi m \qquad m \text{ integral}$$

or equivalently

$$e^{i\underline{k} \cdot \underline{t}_{\underline{k}}} = 1.$$

In our present example for instance

$$\mathfrak{I}_{\Gamma} = \mathfrak{I} = \left\{ \{E \mid n_1\underline{a}_1 + n_2\underline{a}_2 \} \right\}$$

$$\mathfrak{I}_X = \left\{ \{E \mid 2n_1\underline{a}_1 + n_2\underline{a}_2 \} \right\}$$

$$\mathfrak{I}_M = \left\{ \{E \mid 2n_1\underline{a}_1 + 2n_2\underline{a}_2 \} \right\}$$

Since for every element in $\mathfrak{I}_{\underline{k}}$

$$e^{i\underline{k} \cdot \underline{t}_{\underline{k}}} = 1,$$

every element in a particular coset of the factor group $\varkappa/\mathfrak{I}_{\underline{k}}$ yields the same result when it operates on a Bloch function whose index is that $\underline{k}$ which defined $\varkappa$. Thus if $\{R \mid \underline{\tau}\}$ is in $\varkappa$, then a specific coset of $\varkappa/\mathfrak{I}_{\underline{k}}$ is defined by $\{R \mid \underline{\tau}\}\mathfrak{I}_{\underline{k}}$, or, since $\mathfrak{I}_{\underline{k}}$ is invariant, by $\mathfrak{I}_{\underline{k}}\{R \mid \underline{\tau}\}$. A particular element of the coset is $\{E \mid \underline{t}_{\underline{k}}\}\{R \mid \underline{\tau}\}$. Hence

$$\hat{P}_{\{E \mid \underline{t}_{\underline{k}}\}}\hat{P}_{\{R \mid \underline{\tau}\}}\psi_{\underline{k}}(\underline{r}) = \hat{P}_{\{E \mid \underline{t}_{\underline{k}}\}}\psi'_{\underline{k}}(\underline{r}), (\{R \mid \underline{\tau}\} \text{ in } \varkappa)$$

$$= e^{i\underline{k} \cdot \underline{t}_{\underline{k}}} \psi'_{\underline{k}}(\underline{r}) , (\psi'_{\underline{k}} \text{ is Bloch fn.})$$

$$= \psi'_{\underline{k}}(\underline{r}) \qquad , (\{E \mid \underline{t}_{\underline{k}}\} \text{ in } \mathfrak{I}_{\underline{k}})$$

We can generate the whole coset of operations by including all the allowed $\underline{t}_{\underline{k}}$, but the result is independent of $\underline{t}_{\underline{k}}$; so it is the same for all elements of the coset.

The procedure now is to form the factor group $\varkappa/\mathfrak{J}_k$ and to find its irreducible representations. This group generally will be small, but not necessarily iso-morphic with a point group. We build a representation of $\mathfrak{G}$ from a set of Bloch functions ψ_k which transforms under all operations of $\varkappa$ according to an irreducible represen-tation of its factor group $\varkappa/\mathfrak{J}_k$. (This is only possible because every element of a given coset has the same effect.) Then we generate from this set exactly enough functions to span an irreducible representation of $\mathfrak{G}$ by using opera-tions not in $\varkappa$. This irreducible representation is la-beled completely by the k-vector of the generating func-tions and the irreducible representation of $\varkappa/\mathfrak{J}_k$ accord-ing to which they transform under the operations of $\varkappa$. Its essential degeneracy is equal to the dimension of that irreducible representation.

There is, however, one problem with this method: some irreducible representations of $\varkappa/\mathfrak{J}_k$ may not be used to generate an irreducible representation of $\mathfrak{G}$ labeled by the given special k-vector. The reason for this is as follows: Recall that for the given k-vector the invariant subgroup of translations $\mathfrak{J}_k$ was defined by the requirement that $e^{i\underline{k}\cdot\underline{t}} = 1$ for all $\{E \mid \underline{t}\}$ in $\mathfrak{J}_k$. But this require-ment also implies that $(e^{i\underline{k}\cdot\underline{t}})^n = 1^n$, or $e^{i(n\underline{k})\cdot\underline{t}} = 1$. Thus the subgroup $\mathfrak{J}_k$ is appropriate not only to $\underline{k}$ but also to $n\underline{k}$, where n is any integer. Consequently only some of the irreducible representations of $\varkappa/\mathfrak{J}_k$ are appropriate to $\underline{k}$.

Thus in our example, where the k-vector at X is just $\frac{1}{2}\underline{K}_1$, $\mathfrak{J}_k$ is the group of translations $\{E \mid 2n_1\underline{a}_1 + n_2\underline{a}_2\}$ for all integral n_1 and n_2. But $\mathfrak{J}_k$ is appropriate not only to $\underline{k} = \frac{1}{2}\underline{K}_1$ but also to $\underline{K}_1$, $\frac{3}{2}\underline{K}_1$, $2\underline{K}_1$, and so forth. However $\frac{1}{2}\underline{K}_1$, $\frac{3}{2}\underline{K}_1$, $\frac{5}{2}\underline{K}_1$, and so forth, all label the same representation of $\mathfrak{J}$ and so are equivalent. Similarly $\underline{K}_1$, $2\underline{K}_1$, and so on, are all equivalent to one another and also to $\underline{k} = 0$, the special point Γ. And so in this case there will be two types of representation of $\varkappa/\mathfrak{J}_k$. We discrimi-nate against the spurious representations by admitting only those for which all the lattice translations are represented

by $e^{i\underline{k}\cdot\underline{t}}$ multiplied by a unit matrix and not by $e^{in\underline{k}\cdot\underline{t}}$
times the unit matrix.

Thus in our example for X, all the lattice translations not in $\mathfrak{J}_{\underline{k}}$ are of the form

$$\{E \mid (2n_1 + 1)\underline{a}_1 + n_2\underline{a}_2\}.$$

Hence for these translations,

$$e^{i\underline{k}\cdot\underline{t}} = e^{i\frac{1}{2}\underline{K}\cdot[(2n_1 + 1)\underline{a}_1 + n_2\underline{a}_2]} = e^{i[n_1\underline{K}_1\cdot\underline{a}_1 + \frac{1}{2}\underline{K}_1\cdot\underline{a}_1]} = -1.$$

Hence we must keep only representations for which the
translations not in $\mathfrak{J}_k$ are represented by matrices which
are the negative of those representing translations which
are in $\mathfrak{J}_k$.

Now let us follow this program for the point X.
Let us make the following definitions: $\underline{a} = \frac{1}{2}\underline{a}_1$, $\underline{a}_{even} =$
$2n_1\underline{a}_1 + n_2\underline{a}_2$, $\underline{a}_{odd} = (2n_1+1)\underline{a}_1 + n_2\underline{a}_2$, for any n_1 and n_2
which are integers. We easily confirm that χ may be
factored with respect to $\mathfrak{J}_k$ into the following eight
cosets:

$$\{E \mid \underline{a}_{even}\}, \{E \mid \underline{a}_{odd}\}, \{C_{2z} \mid \underline{a}_{even}\}, \{C_{2z} \mid \underline{a}_{odd}\}$$

$$\{m_y \mid \underline{a}+\underline{a}_{even}\}, \{m_y \mid \underline{a}+\underline{a}_{odd}\}, \{m_x \mid \underline{a}+\underline{a}_{even}\}, \{m_x \mid \underline{a}+\underline{a}_{odd}\}.$$

We treat these cosets as elements of a factor group which
we may call $E, \bar{E}, C_{2z}, \bar{C}_{2z}, m_y, \bar{m}_y, m_x, \bar{m}_x$, respectively.
The group multiplication table is easily found and in
this case is isomorphic with the point group 4 mm. Thus
we may write at once the following character table for the
factor group.

	E	$\bar{E}$	$C_{2z}, \bar{C}_{2z}$	$m_y, \bar{m}_y$	$m_x, \bar{m}_x$
	1	1	1	1	1
	1	1	1	-1	-1
	1	1	-1	1	-1
	1	1	-1	-1	1
X_1	2	-2	0	0	0

But since we also have the requirement that $\{E \mid \underline{a}_{odd}\}$ be
represented by the negative of the matrix that represents
$\{E, \underline{a}_{even}\}$, we must discard the first four representations.
The only surviving one is the one labeled X_1, which is two-
dimensional, and so corresponds to a two-fold essential
degeneracy of all energy bands at X. This phenomenon of
having only double levels is known as the "sticking-to-
gether of bands," a terminology which suggests that the
addition of a specific basis to a given Bravais lattice
causes the energy levels to become degenerate (stick-to-
gether) at specific values of $\underline{k}$.

EXAMPLE III: A MODEL CALCULATION OF
ENERGY LEVELS IN A CUBIC CRYSTAL

We turn now to a model calculation which is made
far simpler by applying the results of the theory of symmetry and of groups. As in any quantum mechanical problem,
the central objective of describing a crystalline solid is
attained by diagonalizing its Hamiltonian matrix. The
Hamiltonian operator is characterized, in the one-electron
approximation, by a potential energy with translational symmetry and rotational symmetry fully specified by some particular space-group label; and the Hamiltonian matrix may

be calculated from any suitable complete set of wave functions. But the ease of calculating and diagonalizing the
matrix is almost wholly dependent upon the set chosen, and
any effort expended at this point is amply rewarded later.

For many solids there is abundant evidence that
the important electronic states, the so-called valence and
conduction electrons, are in some general sense very much
like free-electron states. This leads us to hope that the
eigenfunctions of the free-electron Hamiltonian or suitable generalizations might be a good choice as the trial
set. (It is known however that the electronic states are
free-electron-like only far from the ion cores and that
they resemble atomic orbitals in the region of the ion
cores. Our set of functions must then include these characteristics if we expect to obtain a rapid convergence of our
secular equations. Various such sets have been proposed:
the orthogonalized-plane-waves [OPW] and the augmented-
plane-waves [APW] are the best examples. They are an in-
finitely better starting point than the free-electron
plane-wave set. However, since the formal treatment, sym-
metrywise, is very similar, we shall deal only with plane
waves, understanding that all results can be equally applied to OPW's and APW's.)

With the same periodic boundary conditions we have
applied to crystals, the free-electron eigenfunctions are
the well-known plane waves $\frac{1}{\sqrt{\Omega}} e^{i\underline{k}\cdot\underline{X}}$, where Ω is the periodic volume and $\underline{k}$ takes on an unbounded but discrete set of
values which may be represented by points in a k-space
whose separation from one another is of the order of $\frac{2\pi}{L}$,
where L is a characteristic linear dimension of the
periodic volume. In fact, $\hbar\underline{k}$ is the momentum of the state
(and $\frac{\hbar^2 k^2}{2m}$ the energy); so k-space is a momentum space.

Nevertheless it is highly reminiscent of the k-space
of representations of the space groups, and this is not
accidental. Let us look at the symmetry group of the free-
electron Hamiltonian $\frac{-\hbar^2}{am}\boldsymbol{\nabla}^2$. The Laplacian operator is left
invariant by all rotations and all translations. The full
translation group is continuous, and its irreducible rep-

resentations are indeed labeled by the k-vector which
labels the momentum eigenstates. The basic difference,
and a very important one, between the two k-spaces is that
$\underline{k}$ is bounded for crystals and unbounded for free electrons,
allowing an infinite number of irreducible representations
of the continuous translation group in the latter case.

To bring the two situations as nearly into coincide-
dence as possible, we consider a crystal of simple cubic
symmetry without a basis. We apply periodic boundary conditions on a fundamental length L, along the three coordinate axes both to this crystal potential and also to the
vacuum in the free-electron case. This leads, as we saw,
to a densely spaced cubic lattice of points in a $\underline{k}$-space
bounded, in the case of the crystal, by the cubic Brillouin
Zone with faces of $\pm \frac{\pi}{a}$ along the axes. And in the free-
electron case, it leads to the same lattice except that
it is unbounded and there is no Brillouin Zone. Furthermore, both situations have the same point-group symmetry,
namely O_h. (Note that the point-group of the free-electron
Hamiltonian would be the full rotation group, but applying
periodic boundary conditions to a cubic volume of space,
we reduce it to O_h.) Thus the point-group of the two
$\underline{k}$-spaces is also O_h. This means that all states for each
system, which are labeled by a k-vector in the star of a
given $\underline{k}$, are degenerate in the older sense of the word, and
that transform according to the same irreducible representation of the respective space group.

Now what happens when we turn on a weak crystal potential in the cubic system? This problem is precisely the
same as that of the crystal field splitting of atomic levels discussed on pages 94-97 . There we argued that the
eigenfunctions of the unperturbed system, which transformed according to irreducible representations of the
symmetry group of that system, would necessarily transform
according to some, generally reducible, representation of
the group of lower symmetry of the perturbed system, because the latter group is a subgroup of the former. And

so we argue here that the free-electron eigenfunctions
must transform according to some representation of the
crystal space group because the crystal space group
is evidently a subgroup of the continuous space group.
And, as before, a determination of the irreducible components of the representation will give us the splittings
of the free-electron levels.

Consider, for example, the free-electron eigen-
function of k-vector equal to $\frac{2\pi}{a}$ along the x-axis, which
we shall call simply (100). The operations of O_h applied
to $\underline{x}$, or equivalently to $\underline{k}$, generate five other degenerate free-electron eigenfunctions: $(\bar{1}00)$, $(0\underline{1}0)$, $(0\bar{1}0)$,
(001), $(00\bar{1})$. The energy of each is $\frac{h^2(2\pi)^2}{2ma^2}$. But each
of these k-vectors happens to be a reciprocal lattice
vector of the cubic crystal; so each of these wave-func-
tions itself is a periodic function with lattice periodicity. In other words, they must all transform according
to the $\underline{k} = 0$ representations of the crystal space group,
since as Bloch functions they all have the form $\psi_0 =
e^{i0\cdot\underline{x}}u_0(\underline{x}) = u_0(\underline{x})$. The label $\underline{k} = 0$ corresponds to the
point Γ of the Brillouin Zone of the crystal; therefore
we may expect to reduce this representation of its space
group into irreducible components labeled $\Gamma_i^{\pm}$. We may
also expect to find those linear combinations of the six
plane waves which transform according to each irreducible
component (by means of projection operations for example.)

This final step is really our objective, of course,
for it greatly simplifies the calculation and diagonalization of the Hamiltonian matrix. Let us suppose that we
wish to determine the first several energy levels at the
point Γ of our cubic system. The program to follow consists of these steps: (1) List in order of increasing
(unperturbed) energy the (unperturbed) eigenfunctions which,
when put into Bloch form belong to Γ. (2) Collect those
which are transformed into one another by the elements of
the group of Γ and so are degenerate in the unperturbed
system. (3) Determine the characters of the representation of Γ generated by each such set. (4) Find the irre-

ducible components and the appropriate linear combinations of unperturbed eigenfunctions. (5) Using these combinations, construct the Hamiltonian matrix and diagonalize as far as practical.

We carry this out for the point Γ, using the above shorthand labeling for the plane waves: $(rst) \equiv \frac{1}{\sqrt{\Omega}} \exp i(r\underline{K}_1 + s\underline{K}_2 + t\underline{K}_3) \cdot \underline{x}$ where $\underline{K}_i = \frac{2\pi}{a^2} \underline{a}_i$. Here is a list of the first few unperturbed eigenfunctions and their energies:

Unperturbed Energy	0	$\frac{h^2 K^2}{2m}$	$2\frac{h^2 K^2}{2m}$	$3\frac{h^2 K^2}{2m}$	$4\frac{h^2 K^2}{2m}$
Number of Functions	1	6	12	8	6
Functions	(000)	(100)	(110),(1$\bar{1}$0)	(111),(11$\bar{1}$)	(200)
		($\bar{1}$00)	($\bar{1}$10),($\bar{1}\bar{1}$0)	(1$\bar{1}$1),($\bar{1}$11)	($\bar{2}$00)
		(010)	(101),(10$\bar{1}$)	(1$\bar{1}\bar{1}$),($\bar{1}$1$\bar{1}$)	(020)
		(0$\bar{1}$0)	($\bar{1}$01),($\bar{1}$0$\bar{1}$)	($\bar{1}\bar{1}$1),($\bar{1}\bar{1}\bar{1}$)	(0$\bar{2}$0)
		(001)	(011),(01$\bar{1}$)		(002)
		(00$\bar{1}$)	(0$\bar{1}$1),(0$\bar{1}\bar{1}$)		(00$\bar{2}$)

In order to generate the characters of the group of Γ, we let the operations of that group, which is O_h, act upon the k-vectors of all the wave functions in each set and simply count, in the usual way, the number left unaltered by one operation from each class. In this way we get the following table of characters:

	E	$3C_4^2$	$6C_2$	$8C_3$	$6C_4$	i	$3\sigma_v$	$6\sigma_d$	$8S_6$	$6S_4$
(000)set	1	1	1	1	1	1	1	1	1	1
(100)set	6	2	0	0	2	0	4	2	0	0
(110)set	12	0	2	0	0	0	4	2	0	0
(111)set	8	0	0	2	0	0	0	4	0	0
(200)set	6	2	0	0	2	0	4	2	0	0

By referring to the character table for irreducible representations on page 122, we easily confirm that:

$$(000) \rightarrow \Gamma_1^+$$

$$(100) \text{ and } (200) \rightarrow \Gamma_1^+ + \Gamma_3^+ + \Gamma_4^-$$

$$(110) \rightarrow \Gamma_1^+ + \Gamma_3^+ + \Gamma_4^- + \Gamma_5^+ + \Gamma_5^-$$

$$(111) \rightarrow \Gamma_1^+ + \Gamma_2^- + \Gamma_4^- + \Gamma_5^+$$

Finally, we seek the linear combinations of plane waves which transform according to these irreducible components. Thus the six (100)-type functions may be arranged into the following six combinations:

$$\frac{1}{\sqrt{6}}[(100)+(\bar{1}00)+(010)+(0\bar{1}0)+(001)+(00\bar{1})] \rightarrow \Gamma_1^+$$

$$\left. \begin{array}{l} \frac{1}{2}[(100)+(\bar{1}00)-(010)-(0\bar{1}0), \\ \frac{1}{\sqrt{8}}[(100)+(\bar{1}00+(010)+(0\bar{1}0)-2(001)-2(00\bar{1})] \end{array} \right\} \rightarrow \Gamma_3^+$$

$$\frac{1}{\sqrt{2}}[(100)-(\bar{1}00)], \frac{1}{\sqrt{2}}[(010)-(0\bar{1}0)], \frac{1}{\sqrt{2}}[(001)-(00\bar{1})] \rightarrow \Gamma_4^-$$

We find such combinations for each set and use them to construct the Hamiltonian matrix. Now the matrix element theorems developed on pages 114-121 guarantee that every matrix element of H will be zero unless the initial and final states transform according to the same column of the same irreducible representation of the space group of $\hat{H}$. Consequently it is convenient to organize the new wave functions so that all those of a given symmetry fall together: all Γ_1^+; then all Γ_2^+; then all Γ_3^+, column one; then all Γ_3^+ column 2; and so forth. Within each set we may order according to increasing unperturbed energy if we like. In this way the Hamiltonian matrix is brought into block form as far as space-group symmetry will permit. The rest of the problem is actual calculation.

Suppose, for example, that we decide that the lowest several Γ levels can be determined to sufficient accuracy by terminating the series of plane waves after the five sets listed before. This leaves us with 33 wave functions, which are now classified as five Γ_1^+ functions; one Γ_2^- function; three Γ_3^+, column 1, functions; three Γ_3^+, column 2, functions; and so forth. Thus the 33 x 33 matrix is already reduced to a 5 x 5 block, a 1 x 1 block, two equivalent 3 x 3 blocks, and so forth. The four sets of three Γ_4^- functions give three equivalent 4 x 4 blocks; the two sets of three Γ_5^+ functions give three equivalent 2 x 2 blocks; and the single set of three Γ_5^- functions give three equivalent 1 x 1 blocks. Since each of these blocks may be diagonalized independently of the others, the task of splitting an exact diagonalization of this 33 x 33 matrix has been enormously simplified.

A similar analysis can be carried out starting with the sets of plane waves generating representations of the crystal space group - all of which are labeled by another k-vector. However, the symmetry-induced partial diagonalization of the Hamiltonian matrix will be less as the factor group $\kappa/\mathfrak{I}$ of that k becomes smaller. For a general k-vector, the group is merely the identity, and there is no such simplification introduced. Thus it is that for a given accuracy, energy levels are most easily determined at Brillouin Zone points of highest symmetry.

VII

Time-Reversal Symmetry

The idea of time-reversal symmetry is expressed most vividly and succinctly by a comparison with motion picture film. Suppose that all events of the universe were recorded on a motion picture film over some time interval. The principle of time-reversal symmetry says that the succession of events displayed when the film is run backward through the projector is governed by exactly the same deterministic laws which governed the original evolution; and so it is in some sense a "possible" or an "observable" succession of events in its own right.

Thus this idea is closely related to the principle of determinism, which lies at the very heart of physics. The idea of determinism is that a knowledge of the state of things at any instant of time is sufficient to fix once and for all the state at any other time - future and past. Having accepted this principle we must now find the laws governing events. The science of mechanics, whether it be classical, relativistic, or quantum, has always proceeded from this principle. There are three central problems which the inventor of a working mechanics must solve. First, he must decide what constitutes a complete description of the "state" of a system. Second, he must discover the "actions", or "interactions" if the system is isolated, responsible for producing changes of state. Third, he must find a "law of motion" which relates these causes to the resultant changes.

In classical mechanics the state of the system is specified at any time by the values of the positions and velocities of all the component particles, or, from the viewpoint of the Hamiltonian formalism, the values of the

generalized coordinates and momenta of the system. In
quantum mechanics the state is specified by a function of
all the generalized coordinates of the system. In classical mechanics the actions and interactions are merely the
force laws, or, from the other point of view, the Hamiltonian function. In quantum mechanics they are specified
by the Hamiltonian operator. In classical mechanics the
laws of motion are Newton's laws, or Hamilton's equations.
In quantum mechanics Schrödinger's time-dependent equation
is the governing law. (It is perhaps worth emphasizing, in
this context, that the uncertainty principle of quantum
mechanics is embodied in the statistical interpretation
of the state function, but that the evolution of the state
function is completely deterministic.)

Now let us be more specific about the principle of
time-reversal symmetry in the realm of mechanics. Let us
suppose that we have an isolated system whose state is
fully specified at t = 0. According to the idea of mechanics it will subsequently evolve in a manner completely
specified by the nature of the mutual interactions and the
law of motion and will arrive after a time t in a predeter-
mined state. Now let us take a motion picture record of
this evolution of events and run it backward through the
projector. The principle of time-reversal symmetry in
mechanics says that the laws of motion and the interactions
should be such that the reversed succession of events is
the one which would be determined by those laws and interactions and by the initial state of the backward-running
film. The initial state of the backward-running film is
evidently very closely related to the final state of the
normal succession of events, but it is not in general the
exact same state; for if it were, then the laws of motion
and the interactions would simply carry the motion forward
in time from that instant rather than producing the set
of events displayed by the reversed film. We say that
each of these states is the time-reversed conjugate of the
other. Thus each state of the backward-running film is
the time-reversed conjugate of a state of the original
system.

It is easy in classical mechanics to infer explicitly the relationship between time-reversed conjugate states; because at the instant we begin to run the film backward, we see no change in the particle coordinates but we see their velocities reversed in direction. Thus the conjugate state is reached by reversing the particle velocities of the given state. (In the Hamiltonian formalism the generalized momenta are reversed. This is true even when there are magnetic interactions and $\underline{p}_i = m\underline{v}_i + \frac{e}{c}\underline{A}$, because $\underline{A}$ is a linear function of the remaining particle velocities and so itself reverses when they do. It is not true if $\underline{A}$ is an external field.)

At this point we may make contact with physical reality and see that the principle of time-reversal symmetry is not idle speculation. We may take a system in a known state and determine its state at a later time. Then we may take the system in the time-reversed conjugate of that final state and see whether after the same lapse of time its state is the time-reversed conjugate of the original initial state. The fact that experiment has always confirmed this result for isolated systems for which the classical description of the state is expected to be adequate leads one to put great faith in the principle and to use it as a guide in postulating laws of force and laws of motion. Thus, in particular, Newton's second law and Hamilton's equations are manifestly invariant under time-reversal, as are all the well-known interactions of an isolated system.

When we come to the matter of time-reversal symmetry in quantum mechanics there are as many different avenues of approach as there are authors on the subject. One favorite beginning is to define a time-reversal operator $\hat{T}$ which acting on a state ψ yields the time-reversed conjugate state $\psi^{(T)} \equiv \hat{T}\psi$ by the requirements $\hat{T}\hat{x}\hat{T}^{-1} = \hat{x}$, $\hat{T}\hat{p}\hat{T}^{-1} = -\hat{p}$, and $\hat{T}\hat{\sigma}\hat{T}^{-1} = -\hat{\sigma}$. This is a physically appealing approach, especially since it offers a definition of $\hat{T}$ which is invariant under changes of representation. However, it does require our faith that the spin degree of freedom is so much like ordinary angular momentum that it transforms like it under time reversal, and such may not be the case.

The following line of argument seems more satisfactory to us. Pursuing the motion picture film idea and relying on our faith in the principle of time reversal symmetry, we begin by requiring that our mechanics exhibit this symmetry and then inquire into the nature of the time-reversed state. We assert that the successive application of the following three operations to an arbitrary state function $\psi(\underline{r})$ must yield its time-reversed conjugate $\hat{T}\psi$: first, an evolution in time by an arbitrary amount t; second, a time reversal; third, another evolution in time by the same amount t. Thus we say, (evolution by t) x (time reversal) x (evolution by t)ψ = (time reversal)ψ. The time evolution is governed by Schrödinger's equation

$$i\hbar\frac{\partial\psi}{\partial t} = \hat{H}\psi$$

which may be integrated formally to get the time-evolution operator $e^{-\frac{i}{\hbar}\hat{H}t}$ such that

$$\psi(\underline{r},\underline{t}) = e^{-\frac{i}{\hbar}\hat{H}t}\psi(\underline{r}).$$

Since the equation above must hold for all state functions, we are led to the following operator equation:

$$e^{-\frac{i}{\hbar}\hat{H}t}\hat{T}\,e^{-\frac{i}{\hbar}\hat{H}t} = \hat{T}$$

We multiply from the left by $\hat{T}^{-1}$ and from the right by $e^{+\frac{i}{\hbar}\hat{H}t}$ and get

$$\hat{T}^{-1}e^{-\frac{i}{\hbar}\hat{H}t}\hat{T} = e^{\frac{i}{\hbar}\hat{H}t} \quad\text{or}\quad e^{-\hat{T}^{-1}\frac{i}{\hbar}\hat{H}t\hat{T}} = e^{\frac{i}{\hbar}\hat{H}t}.$$

Clearly the time-reversal operator must be independent of the value of the time, which from the point of view of mechanics is merely a parameter and not directly involved in a description of the state of the system. Thus

$\hat{T}t = t\hat{T}$, and we may write

$$e^{-\hat{T}i\hat{H}\hat{T}^{-1}\frac{t}{\hbar}} = e^{i\hat{H}\frac{t}{\hbar}}$$

This operator equation must be true for any interval of time t; therefore we must equate all powers of t in the series which the exponentials are understood to stand for. This leads to a set of equations that may be satisfied only if

$$\hat{T}i\hat{H}\hat{T}^{-1} = -i\hat{H}$$

Now let us recall that although the law of motion of quantum mechanics is Schrödinger's equation, or equivalently the time-evolution operator, the nature of the forces or interactions is only specified by the Hamiltonian operator. And in setting forth the principle of time-reversal symmetry we specifically required that the same interactions be applied to the time-reversal state. Thus the Hamiltonian must itself be invariant under time reversal if the system is to exhibit time reversal symmetry: $\hat{T}\hat{H}\hat{T}^{-1} = \hat{H}$. And we are left finally with the result that in order for a system whose interactions are governed by a time-reversal invariant Hamiltonian and whose evolution is governed by Schrödinger's equation to exhibit time-reversal symmetry it is necessary that the time-reversal operator satisfy the condition

$$\hat{T}i\hat{T}^{-1} = -i \quad \text{or} \quad \hat{T}i = -i\hat{T}.$$

The important conclusion here is that $\hat{T}$ is not a linear operator:

$$\hat{T}(a\psi + b\Phi) = a^*\hat{T}\psi + b^*\hat{T}\Phi \neq a\hat{T}\psi + b\hat{T}\Phi.$$

Such an operator is called anti-linear. We show now that it is also anti-unitary:

$$(\hat{T}\psi, \hat{T}\Phi) = (\psi, \Phi)^* = (\Phi, \psi)$$

We invoke only the very reasonable assumption that time reversal does not affect normalization or orthogonality of state functions. Let us expand ψ and Φ in terms of the ortho-normal eigenfunctions of $\hat{H}$, so that

$$(\psi, \Phi) = \sum_{n,m}(a_n\psi_n, b_m\psi_m) = \sum_{n,m} a_n{}^* b_m \delta_{nm} = \sum_n a_n{}^* b_n$$

where $\hat{H}\psi_n = E_n\psi_n$.

We may also write

$$(\hat{T}\psi, \hat{T}\Phi) = \sum_{n,m}(\hat{T}a_n\psi_n, \hat{T}b_m\psi_m) = \sum_{n,m}(a_n{}^*\hat{T}\psi_n, b_m{}^*\hat{T}\psi_m)$$

$$= \sum_{n,m} a_n b_m{}^*(\hat{T}\psi_n, \hat{T}\psi_m) = \sum_{n,m} a_n b_m{}^* \delta_{nm}$$

$$= \sum_n a_n b_n{}^* = (\psi, \Phi)^*, \quad \text{Q.E.D.}$$

The underlying assumption is made all the more reasonable when we note that the set $\{\hat{T}\psi_n\}$ are also eigenfunctions of $\hat{H}$ and are degenerate one-to-one with the set $\{\psi_n\}$. For

$$\hat{T}\hat{H}\psi_n = \hat{T}E_n\psi_n \rightarrow \hat{H}\hat{T}\psi_n = E_n\hat{T}\psi_n, \text{ since } E_n \text{ is real.}$$

The operation of complex conjugation $\hat{K}$ is easily seen to be antilinear and antiunitary, and it is in a sense the generator of all antiunitary antilinear operators. Note that the product of two antilinear operators, such as $\hat{T}\hat{K}$, is linear:

$$\hat{T}\hat{K}(a\psi + b\Phi) = \hat{T}(a^*\hat{K}\psi + b^*\hat{K}\Phi) = a\hat{T}\hat{K}\psi + b\hat{T}\hat{K}\Phi.$$

And the product is also unitary:

$$(\hat{T}\hat{K}\psi, \hat{T}\hat{K}\Phi) = (\hat{T}\psi, \hat{T}\Phi)^* = (\psi, \Phi).$$

Thus if we write $\hat{T}\hat{K} = \hat{U}$, where $\hat{U}$ is some unitary operation, and multiply on the right by $\hat{K}$, noting that $\hat{K}^2 = 1$, we get $\hat{T} = \hat{U}\hat{K}$. That is any antiunitary operator may be factored into the product of some unitary operator and complex conjugation.

Before going on to an explicit determination of $\hat{T}$ we consider one final property. Two successive applications of time reversal to an arbitrary state must above all return the system to a state in no way distinguishable from its original state. Thus the only permissible effect of the double application would be a change of phase. Hence we may write $\hat{T}^2 = c1$ where $|c| = 1$. Let us write $\hat{T} = \hat{U}\hat{K}$ and see what conditions this places upon $\hat{U}$.

$$\hat{U}\hat{K}\hat{U}\hat{K} = c1$$

$$\hat{U}\hat{U}^*\hat{K}\hat{K} = c1 \quad (\text{definition of } \hat{K})$$

$$\hat{U}\hat{U}^* = c1 \quad (\hat{K}^2 = 1)$$

$$\hat{U}^* = c\hat{U}^\dagger \quad (\hat{U}\hat{U}^\dagger = 1)$$

$$\hat{U} = c\hat{\tilde{U}} \quad (\text{operate on left with } \hat{K})$$

$$\text{but} \quad \hat{\tilde{U}} = c\hat{U} \quad (\text{take transpose})$$

$$\text{hence} \quad \hat{U} = c^2\hat{U} \quad (\text{substitution})$$

$$\text{and} \quad c = \pm 1$$

And so we see that $\hat{U}$ is either completely symmetric or else completely antisymmetric. Note that we cannot remove the case $c = -1$, should it occur, simply by redefining $\hat{T}$ to include some constant phase factor. For if $\hat{T}^2 = -1$ and $\hat{T}' = a\hat{T}$, then $\hat{T}'^2 = a\hat{T}a\hat{T} = aa^*\hat{T}^2 = -|a|^2 = -1$

<u>Determination of $\hat{T}$.</u> To write down an explicit form for $\hat{T}$ we must, at last, make some requirements on the observables of the time-reversed state. There are two important classes of physical observables: those, like position and energy, which in classical mechanics are associated with an even power of the time parameter, and those, like velocity, momentum, and angular momentum, associated with an odd power of time. Now we require that the probability for a specific value of an observable be the same in the original and the time-reversed states if it is of

the first class, while it be of opposite sign if the observable is in the second class. Thus if $\hat{O}$ is the operator corresponding to one of these observables, we say that $(\hat{T}\psi, \hat{O}\hat{T}\psi) = \pm(\psi, \hat{O}\psi)$ where the signs correspond to the two classes. Thus we may write

$$\pm(\psi, \hat{O}\psi) = (\hat{T}\psi, \hat{T}\hat{T}^{-1}\hat{O}\hat{T}\psi)$$

$$= (\psi, \hat{T}^{-1}\hat{O}\hat{T}\psi)^* \quad \text{(by antiunitarity of } \hat{T})$$

$$\pm(\psi, \hat{O}\psi)^* = (\psi, \hat{T}^{-1}\hat{O}\hat{T}\psi) \quad \text{(complex conjugation)}$$

$$\pm(\psi, \hat{O}\psi) = (\psi, \hat{T}^{-1}\hat{O}\hat{T}\psi) \quad \text{(reality of observables)}.$$

Since the result must hold for arbitrary ψ we have the result

$$\hat{T}\hat{O} = \pm \hat{O}\hat{T}$$

Let us now write $\hat{T} = \hat{U}\hat{K}$.

$$\hat{U}\hat{K}\hat{O} = \pm \hat{O}\hat{U}\hat{K}$$

$$\hat{U}\hat{O}^*\hat{K} = \pm \hat{O}\hat{U}\hat{K}$$

$$\text{or} \quad \hat{U}\hat{O}^* = \pm \hat{O}\hat{U}$$

In particular, if $\hat{O} = \underline{\hat{x}}$ we take the upper sign, and since $\underline{\hat{x}}$ is real in the Schrödinger coordinate representation, we get

$$\hat{U}\underline{\hat{x}} = \underline{\hat{x}}\hat{U}$$

If now we put $\hat{O} = \underline{\hat{p}}$, we take the lower sign and since in this representation $\hat{p}$ is imaginary, we get

$$\hat{U}\underline{\hat{p}} = \underline{\hat{p}}\hat{U} \quad \text{or} \quad \hat{U}\underline{\hat{v}} = \underline{\hat{v}}\hat{U}$$

since $\hat{U}$ commutes with i, being a linear operator. Now since U commutes both with $\underline{x}$ and $\underline{v}$ it cannot be a function of x or a differential operator of the coordinates. Thus if there were no spin coordinates it could only be a constant multiplier of modulus unity; and since $\hat{U}$ is only determined up to a phase factor, we can choose $\hat{U} = 1$ and get $\hat{T} = \hat{K}$. (This is only true of the <u>coordinate</u> representation. In the momentum representation U takes $\underline{p}$ into $-\underline{p}$.)

But what if we include the spin coordinates of the
system? We might be tempted to pursue the analogy with
ordinary angular momentum and require at once that $\hat{T}\hat{\sigma} = -\hat{\sigma}\hat{T}$.
It seems logically sounder to proceed from the requirement
that $\hat{T}^{-1}\hat{H}\hat{T} = \hat{H}$ and then to recall that the Hamiltonian in
which spin first enters and which describes nature very
well contains a term of the form[const. $\hat{\underline{\sigma}} \cdot \underline{\nabla}V(\underline{x}) \times \hat{\underline{p}}$]
where the constant is real. Now since $\hat{\underline{p}}$ anticommutes with
$\hat{T}$ it is necessary that $\hat{\sigma}$ also anticommute in order that the
entire term commute. In any event, we are led to the re-
quirement $\hat{U}\hat{\underline{\sigma}}^* = -\hat{\underline{\sigma}}\hat{U}$ or

$$\hat{U}\hat{\sigma}_x = -\hat{\sigma}_x\hat{U}$$
$$\hat{U}\hat{\sigma}_y = \hat{\sigma}_y\hat{U}$$
$$\hat{U}\hat{\sigma}_z = -\hat{\sigma}_z\hat{U}$$

This is true in the usual representation of the Pauli
matrices where $\hat{\sigma}_y$ is imaginary and the others are real.
Since the 2 x 2 unit matrix and the three Pauli matrices
completely span the space of 2 x 2 matrices, it is easy
to confirm that, apart from an arbitrary phase factor,
$\hat{U} = \hat{\sigma}_y$ and $\hat{T} = \hat{\sigma}_y K$. Note that here $\hat{T}^2 = \hat{\sigma}_y K \hat{\sigma}_y K = -\hat{\sigma}_y\hat{\sigma}_y KK = -1$,
while it was equal to + 1 in the spinless case.

If the system contained N particles we evidently
would have found the result

$$\hat{U} = \prod_{i=1}^{N} \hat{\sigma}_{yi}$$

$$\hat{T} = \hat{U}\hat{K}$$

In this case $\hat{T}^2 = (-1)^N$.

<u>Time-Reversal Symmetry and Degeneracy</u>. Since $\hat{T}$ is
a symmetry operator of the Hamiltonian $(\hat{T}\hat{H}\hat{T}^{-1} = \hat{H})$, we can
expect its presence to require additional degeneracies
among the energy eigenfunctions. For if ψ_n is such that

$$\hat{H}\psi_n = E_n\psi_n, \text{ then}$$
$$\hat{T}\hat{H}\psi_n = \hat{T}E_n\psi_n \text{ or}$$
$$\hat{H}(\hat{T}\psi_n) = E_n(\hat{T}\psi_n)$$

so that the time-reversed state $\hat{T}\psi_n$ is also an eigen-function of $\hat{H}$ degenerate with ψ_n, provided that it is indeed linearly independent of ψ_n.

Suppose, in particular, that the set of functions $\{\psi_\mu^{(j)}\} \equiv \underline{\psi}^{(j)}$ is required by the space symmetries already discussed to be degenerate and in fact transform according to the $j^{\underline{th}}$ irreducible representation of the group $\mathcal{G}$ of such operations. Then, by the argument previously given, the set $\hat{T}\underline{\psi}^{(j)}$ must be degenerate with the set $\underline{\psi}^{(j)}$ provided once again that the new set is linearly independent of the old set.

There are two avenues open to us now. We can search for simple criteria to tell us when the sets are linearly independent, or we can try to absorb $\hat{T}$ into the group of operations as though it were another space symmetry operation. The latter approach is certainly more general and more likely to supply us with useful theorems.

Suppose the set $\underline{\psi}$ are ortho-normal basis vectors of a subspace which is invariant under $\mathcal{G}$ and also under $\hat{T}$. Now we enlarge $\mathcal{G}$ by adding all the elements $\hat{T}\mathcal{G}$. (The result is also a group if $\hat{T}^2 = 1$ since $\hat{T}$ commutes with all the elements of $\mathcal{G}$ and $\hat{T}$ and $\hat{P}_E$ form a group. It is in fact a direct product group. If $\hat{T}^2 = -1$ and $\mathcal{G}$ is taken to mean the <u>double</u> group then once more the result is a group.) The operations of $\mathcal{G}$ are all unitary, while those of $\hat{T}\mathcal{G}$ must be antiunitary. Now let us form the representation matrices of these operations generated by the set $\underline{\psi}$. Since antiunitary, like unitary, operations preserve normalization, the earlier proof on page 54 of the unitarity of the matrices goes through in the same way.

Now let us look at the representation of the product

of two operations $\hat{P}_R \hat{P}_S$.

$$\hat{P}_R \hat{P}_S \underline{\psi} = \underline{\psi}\Gamma(RS), \quad \text{but also}$$

$$\hat{P}_R(\hat{P}_S\underline{\psi}) = \hat{P}_R\underline{\psi}\Gamma(S) = \begin{cases} \underline{\psi}\Gamma(R)\Gamma(S) & \text{if } \hat{P}_R \text{ linear} \\ \underline{\psi}\Gamma(R)\Gamma^*(S) & \text{if } \hat{P}_R \text{ antilinear} \end{cases}$$

Thus $\Gamma(RS) = \Gamma(R)\Gamma(S)$ only for the case $\hat{P}_R$ unitary, and the matrices do not represent, in the usual sense, the group. To proceed further along these lines would require a suitable generalization of the representation theory which we have developed; so we turn to the other avenue of approach and seek criteria to determine whether the sets $\underline{\psi}^{(j)}$ and some $\hat{T}\underline{\psi}^{(j)}$ are linearly independent.

As a start we calculate formally the inner product of some $\psi_\mu^{(j)}$ and some $\hat{T}\psi_\nu^{(j)}$ where $\psi_\mu^{(j)}$ and $\psi_\nu^{(j)}$ are ortho-normal basis vectors of a vector space invariant under $\mathcal{G}$ and transforming according to $\Gamma^{(j)}$. First we prove that the set $\hat{T}\underline{\psi}^{(j)}$ also are ortho-normal basis vectors of an invariant irreducible subspace.

$$\hat{P}_R(\hat{T}\underline{\psi}) = \hat{T}\hat{P}_R\underline{\psi} \qquad \begin{array}{l}(\hat{T} \text{ commutes with coordinate trans-}\\ \text{formations})\end{array}$$

$$= \hat{T}(\underline{\psi}\Gamma^{(j)}(R)) \qquad \begin{array}{l}(\underline{\psi} \text{ are basis functions of invar-}\\ \text{iant subspace})\end{array}$$

$$= (\hat{T}\underline{\psi})\Gamma^{(j)*}(R) \quad (\hat{T} \text{ antilinear})$$

Thus the set $\hat{T}\underline{\psi}$ is always transformed into itself; so the subspace is invariant. The orthonormality is preserved under $\hat{T}$. To prove the irreducibility we assume the contrary and suppose some U exists such that $\Gamma'(R) = U\Gamma^{(j)}(R)U^{-1}$ is in block form for all R in $\mathcal{G}$. Since complex conjugation leaves the block form unaltered, $\Gamma'^*(R) = U^*\Gamma^{(j)}(R)U^{*-1}$ is also in block form and we have found a matrix U^* which reduces $\Gamma^{(j)}(R)$. But this is contrary to hypothesis, and our original supposition is wrong. Thus $\Gamma^{(j)*}(R)$ for all R is an irreducible representation of $\mathcal{G}$ of the same dimension as $\Gamma^{(j)}(R)$. Now let us carry out the aforementioned inner product $(\psi_\mu^{(j)}, \hat{T}\psi_\nu^{(j)})$.

$$(\psi_\mu^{(j)}, \hat{T}\psi_\nu^{(j)}) = (\hat{P}_R \psi_\mu^{(j)}, \hat{P}_R \hat{T}\psi_\mu^{(j)}) \quad \text{(unitarity of } \hat{P}_R)$$

$$= \sum_{\kappa,\lambda} (\psi_\kappa \Gamma^{(j)}(R)_{\kappa\mu} \hat{T}\psi_\lambda^{(j)} \Gamma^{(j)*}(R)_{\lambda\nu} \quad \text{(previous result)}$$

$$= \sum_{\kappa,\lambda} \Gamma^{(j)*}(R)_{\kappa\mu} (\psi_\kappa^{(j)}, \hat{T}\psi_\lambda^{(j)}) \Gamma^{(j)*}(R)_{\lambda\nu}$$

We now use the fact that $\Gamma^{(j)*}(R)_{\lambda\nu}$ is an irreducible representation and replace it by $\Gamma^{(i)}(R)_{\lambda\nu}$. We sum both sides over R and divide by h using the great orthogonality theorem, page 29 , to get

$$(\psi_\mu^{(j)}, \hat{T}\psi_\nu^{(j)}) = \sum_{\kappa,\lambda} \frac{1}{\ell_j} \delta_{ji} \delta_{\kappa\lambda} \delta_{\mu\nu} (\psi_\kappa^{(j)}, \hat{T}\psi_\lambda^{(j)})$$

$$= \frac{\delta_{ji}\delta_{\mu\nu}}{\ell_j} \sum_{\lambda=1}^{\ell_j} (\psi_\lambda^{(j)}, \hat{T}\psi_\lambda^{(j)})$$

Thus we see at once that if $\Gamma^{(j)*}$ is <u>not</u> equivalent to $\Gamma^{(j)}$ -that is, $i \neq j$- then the entire subspace spanned by $\psi^{(j)}$ is orthogonal to that spanned by $\hat{T}\psi^{(j)}$; so the time-reversed functions must be linearly independent of the original ones, and are, of course, degenerate with them.

If $\Gamma^{(j)*}$ <u>is</u> equivalent to $\Gamma^{(j)}$, there are only two significant possibilities: either $\Gamma^{(j)}$ is or is not equivalent to a <u>real</u> representation. In other words, either the basis functions can or cannot be chosen so that $\Gamma^{(j)}$ is real. If the basis functions can be so chosen, then $\Gamma^{(j)*}$ is <u>identical</u> with $\Gamma^{(j)}$ and our previous result becomes

$$(\psi_\mu^{(j)}, \hat{T}\psi_\nu^{(j)}) = \frac{\delta_{\mu\nu}}{\ell_j} \sum_{\lambda=1}^{\ell_j} (\psi_\lambda^{(j)}, \hat{T}\psi_\lambda^{(j)}) \qquad (*)$$

when the $\psi_\mu^{(j)}$ are so chosen. This tells us that each of the basis functions is orthogonal to all time-reversed basis functions except perhaps to its own time-reverse, and that $(\psi_\mu^{(j)}, \hat{T}\psi_\mu^{(j)})$ is the same for all μ.

Now let us suppose that $\hat{T}\psi^{(j)}$ _is_ linearly dependent upon the set $\{\psi_\mu^{(j)}\}$: $\hat{T}\psi_\mu^{(j)} = \sum_\nu \psi_\nu^{(j)} T_{\nu\mu}$. The matrix of coefficients $T_{\nu\mu}$ must be __unitary__ if normalization is to be preserved. Now let us operate on this equation, noting that $\hat{T}^2 = \pm 1$.

$$\hat{T}^2 \psi_\mu^{(j)} = \sum_\nu \hat{T}(\psi_\nu^{(j)} T_{\nu\mu}) = \sum_{\nu,\lambda} \psi_\lambda^{(j)} T_{\lambda\nu} T_{\nu\mu}^* ; \text{ so}$$

$$\pm 1 = TT^* \text{ or } \pm T^\dagger = T^* \text{ or } \boxed{T = \pm \tilde{T}.}$$

Now we have found that $(\psi_\mu^{(j)}, \hat{T}\psi_\nu^{(j)}) = a\,\delta_{\mu\nu}$; so

$$a\,\delta_{\mu\nu} = \sum_\lambda (\psi_\mu^{(j)}, \psi_\lambda^{(j)}) T_{\lambda\nu} = T_{\mu\nu}$$

Thus T is a constant times the unit matrix and so can be anti-symmetric $(\tilde{T} = -T)$ only if $a = 0$. But this implies that the set $T\underline{\psi}^{(j)}$ is orthogonal to the set $\underline{\psi}^{(j)}$, and our supposition of linear dependence is violated. We must conclude that if $\hat{T}^2 = -1$ the time-reversed states are again linearly independent of the original states and degenerate with them.

But what if $\hat{T}^2 = +1$? For this case we return to equation (*) on page 194 and replace $\hat{T}\psi_\lambda^{(j)}$ on the right side by $\sum_\varkappa \psi_\varkappa^{(j)} T_{\varkappa\lambda}$:

$$(\psi_\mu^{(j)}, \hat{T}\psi_\nu^{(j)}) = \frac{\delta_{\mu\nu}}{\ell_j} \sum_{\lambda,\varkappa} (\psi_\lambda^{(j)}, \psi_\varkappa^{(j)}) T_{\varkappa\lambda} = \frac{\delta_{\mu\nu}}{\ell_j} \sum_\lambda T_{\lambda\lambda} \neq 0$$

where the last result follows from the fact that T is proportional to the unit matrix.

Actually this only proves that the original assumption of linear dependency leads to no inconsistency in this case. It is easy to show, however, that it actually occurs. If the set $\{\psi_\mu^{(j)}\}$ has been chosen so that $\Gamma^{(j)}(R)$ is real, then the set $\{\hat{T}\psi_\mu^{(j)}\}$ generates exactly the same representation and so will any linear combination $\{a\psi_\mu^{(j)} + b\hat{T}\psi_\mu^{(j)}\}$. In particular we form a new set $\{\phi_\mu^{(j)}\}$ with $\phi_\mu^{(j)} = (\psi_\mu^{(j)} + \hat{T}\psi_\mu^{(j)})n$ where n is a real normalizing factor, and we assume that $\hat{T}\psi_\mu^{(j)} \neq -\psi_\mu^{(j)}$. If the last condition is not fulfilled for some μ, we take $\phi_\mu^{(j)} = i(\psi_\mu^{(j)} - \hat{T}\psi_\mu^{(j)})n$ instead. Now it is easy to confirm in either case that if $\hat{T}^2 = +1$, then $\hat{T}\phi_\mu^{(j)} = \phi_\mu^{(j)}$; so the time-reversed states $\{\hat{T}\phi_\mu^{(j)}\}$ are linearly dependent on the original set $\{\phi_\mu^{(j)}\}$. And of course this fact must remain under any unitary transformations performed on the $\{\phi_\mu^{(j)}\}$. Thus there can be no new degeneracy here.

Finally we return to the case where $\Gamma^{(j)}(R)^*$ is equivalent to $\Gamma^{(j)}(R)$ but $\Gamma^{(j)}(R)$ cannot be made real by any unitary transformation of the basis functions. It follows that no matter how the basis functions are chosen, the sets $\{\psi_\mu^{(j)}\}$ and $\{\hat{T}\psi_\mu^{(j)}\}$ must generate distinct, although equivalent, representations, and our previous method of approach breaks down. A rigorous analysis of this case is very tedious, and we simply state here the result that extra degenerate functions are implied only in the case that $\hat{T}^2 = +1$.

With these results we have a complete set of criteria for determining additional degeneracies brought about through time-reversal symmetry. Nevertheless it may still appear less than obvious which of the three cases we are dealing with, especially if we have only the character table of a given group at our disposal. If the group is not of unreasonably high order, the Frobenius-Schur test is a simple one to apply to this question. For convenience we enumerate the cases here.

Case a: Γ and Γ^* are equivalent and can be chosen to be real and identical.

Case b: Γ and Γ^* are inequivalent.

Case c: Γ and Γ^* are equivalent but cannot be chosen to be real and identical.

The Frobenius-Schur test requires us to find the character of the square of each element R in $\mathfrak{G}$ and to sum the

characters over all R in $\mathcal{G}$. If the sum is h, the order
of $\mathcal{G}$, we have Case a. If it is zero, we have Case b.
If it is -h we have Case c.

Proof: $\chi^{(j)}(R^2) = \mathrm{Tr}\, \Gamma^{(j)}(R^2) = \mathrm{Tr}[\Gamma^{(j)}(R)\Gamma^{(j)}(R)]$

We write this with indexes, sum over R, and use the great
orthogonality theorem, page 29.

$$\sum_R \chi^{(j)}(R^2) = \sum_{R}\sum_{\mu,\nu} \Gamma^{(j)}(R)_{\mu\nu}\Gamma^{(j)}(R)_{\nu\mu}$$

Case a: Since $\chi(R)$ is the same for equivalent representations, we may assume that $\Gamma^{(j)}(R)$ has been made
real already. The orthogonality theorem can be applied:

$$\sum_R \chi^{(j)}(R^2) = \sum_{\mu,\nu} \frac{h}{\ell_j}\delta_{\mu\nu}\delta_{\nu\mu} = \sum_{\mu}\frac{h}{\ell_j}\delta_{\mu\mu} = h$$

Case b: $\Gamma^{(j)*}(R) = \Gamma^{(k)}(R), \ k \neq j$

$$\sum_R \chi^{(j)}(R^2) = \mathrm{const.} \times \delta_{jk} = 0$$

Case c: $\Gamma^{(j)*}(R) = U\Gamma^{(j)}(R)U^\dagger$

$$\sum_R \chi^{(j)}(R^2) = \sum_{R}\sum_{\substack{\mu,\nu\\ \varkappa,\lambda}} \Gamma^{(j)}(R)_{\mu\nu}U^*_{\nu\varkappa}\Gamma^{(j)*}(R)_{\varkappa\lambda}U_{\mu\lambda}$$

$$= \sum_{\mu,\nu}\frac{h}{\ell_j}U^*_{\nu\mu}U_{\mu\nu} = \frac{h}{\ell_j}\,\mathrm{Tr}(U^*U)$$

It can be shown that U must be anti-symmetric if $\Gamma^{(j)}(R)$
cannot be made real. Thus $U^* = -U^\dagger = -U^{-1}$, $\sum_R \chi^{(j)}(R^2) = -h$.

Q.E.D.

It is worth noting that in performing the sum over R it
is sufficient to know only one R^2 and one $\chi(R^2)$ from

each class. For if R and S belong to the same class, that
is, $TRT^{-1} = S$ for some T in $\mathcal{G}$, then R^2 belongs to the class
of S^2. $S^2 = TRT^{-1}TRT^{-1} = TR^2T^{-1}$. And so $\chi(S^2) = \chi(R^2)$.
Thus we may write

$$\sum_R \chi(R^2) = \sum_k N_k \chi(R_k^2)$$

where N_k is the number of elements in class C_k and where
R_k is any single element in C_k.

Note: If $\mathcal{G}$ is a double group, we must be careful
about squaring elements. For example, $(C_2)^2 = \overline{E}$.

Time reversal and space groups. Since space groups
are virtually infinite in order, the application of the
Schur-Frobenius test is impractical in its present form.
But Herring has shown that it is sufficient to limit the
sum to some elements whose squares are in the factor group
$\mathcal{K}/\mathcal{J}$ of the group of the k-vector, $\mathcal{K}$. The elements Q_0 are
those which take $\underline{k}$ into $-\underline{k}$; so Q_0^2 must be in $\mathcal{K}$. If there
are n such elements, the Herring test is

$$\sum_{Q_0} \chi(Q_0^2) = \begin{cases} n - \text{case a} \\ 0 - \text{case b} \\ -n - \text{case c} \end{cases}$$

Note: $\chi(Q_0^2)$ is the character of Q_0^2 in the group of $\mathcal{K}$
not in the space group $\mathcal{G}$.

The following rules are very helpful:

(1) If $\mathcal{G}$ contains the inversion i and $\mathcal{K}$ also contains i, then the Q_0 are the elements of $\mathcal{K}$.

(2) If $\mathcal{G}$ contains i but $\mathcal{K}$ does not, then the Q_0
are the elements $i\mathcal{K}$.

In the following table we summarize all the results
from our study of time-reversal symmetry.

Case	a	b	c
Relation of Γ to Γ^*	Can be made real and identical	Inequivalent	Equivalent but cannot be made real
Frobenius - Schur Test	h	0	-h
Herring Test	n	0	-n
Degeneracies for Integral Spin	none	doubled	doubled
Degeneracies for Half-Integral Spin	doubled	doubled	none

VIII

Continuous Groups

The Axial Rotation Group and the
Full Rotation Group

CONTINUOUS GROUPS

In our discussion of the representations of groups
we have restricted ourselves to groups with a finite number h of elements. This condition was necessary to prove
several properties, the most important one being the rearrangement theorem which established that, when summing
of all elements R of the group, we can replace R by SR,
where S is a fixed given element. In other words, if R
runs over the h elements of the group, so does SR. Consequently,

$$\sum_R D(R) = \sum_R D(SR)$$

where $D(R)$ is any quantity which depends on R.

Here we shall be concerned with infinite <u>continuous
groups</u>, that is, groups which have an infinite number of
elements, each of which may be labeled by n real parameters
which vary continuously.

$$R(p_1, p_2, p_3, \ldots, p_n) = R(\underline{p})$$

The set of parameters $\underline{p} = p_1, p_2, p_3, \ldots, p_n$ can be considered an n-dimensional vector which may vary continuously
in a given domain. We shall also be concerned with groups
in which $\underline{p}$ varies continuously within each one of a finite
number of regions, but discontinuously from region to region. Such groups are called <u>mixed continuous groups</u>.

The continuous groups satisfy all four group properties, and the group multiplication

$$R(\underline{q}) \ R(\underline{p}) = R(\underline{t})$$

is given in this case by a set of n real functions

$$t_1 = f_1(p_1, \ p_2 \ \cdots \ p_n, \ q_1 \ \cdots \ q_n)$$

$$t_2 = f_2(p_1, \ p_2 \ \cdots \ p_n, \ q_1 \ \cdots \ q_n)$$

$$\vdots$$

$$t_n = f_n(p_1, \ p_2 \ \cdots \ p_n, \ q_1 \ \cdots \ q_n)$$

which give the vector $\underline{t}$ in terms of the vectors $\underline{p}$ (applied first) and $\underline{q}$ (applied second). The functions $\underline{t} = f(\underline{p}, \ \underline{q})$ are such that they guarantee associativity, existence of a unit element, and existence of the inverse element. If, in addition to being continuous, these functions possess derivatives of all orders, we call the group a Lie group.

Examples: (A) The group of linear transformations in one dimension.

$$x' = a_1 x + a_2 \qquad a_1 \neq 0$$

It is a two-parameter group.

"Multiplication rule":

$$\begin{cases} x' = a_1 x + a_2 \\ x'' = b_1 x' + b_2 \\ x'' = c_1 x + c_2 \end{cases}$$

$$\therefore \ c_1 = b_1 a_1 \qquad c_2 = b_1 a_2 + b_2$$

Unit element : $\quad x = e_1 x + e_2$

$$e_1 = 1 \qquad e_2 = 0$$

Inverse element:
$$x' = a_1 x + a_2$$
$$x = \bar{a}_1 x' + \bar{a}_2$$

$$\bar{a}_1 = \frac{1}{a_1} \qquad \bar{a}_2 = \frac{-a_2}{a_1}$$

It is a non-abelian group.

(B) The group of rotations about the z-axis.
(<u>Axial rotation group</u>)

$$x' = \cos\varphi \; x + \sin\varphi \; y$$
$$y' = -\sin\varphi \; x + \cos\varphi \; y. \qquad 0 \leqslant \varphi < 2\pi$$

It is a one-parameter group.

"Multiplication rule": $\qquad \varphi_c = \varphi_a + \varphi_b$

Unit element : $\qquad \varphi_e = 0$

Inverse element : $\qquad \varphi_{-a} = -\varphi_a$

It is an abelian group.

(C) The groups SU(n) of all n x n unitary matrices with determinant 1. It has n^2-1 parameters. In particular an element of SU(2) can be written in terms of the three parameters ξ, η, ζ:

$$\begin{pmatrix} e^{i\xi}\cos\eta & e^{i\zeta}\sin\eta \\ -e^{-i\zeta}\sin\eta & e^{-i\xi}\cos\eta \end{pmatrix}$$

(D) The group O(n) of all real orthogonal n x n ma-trices. It is a mixed continuous group since those matri-ces with determinant -1 cannot be reached by a continuous variation of the parameters starting from one matrix of determinant +1. If we restrict the matrices to those of determinant +1, the group is denoted by $O^+(n)$. In particu-lar, $O^+(3)$ is isomorphic with the group of proper rotations in three dimensions (full rotation group) and $O(3)$ is isomorphic with the group of proper and improper rotations in three dimensions (rotation-inversion group). $O(3)$ is a three-parameter group and is homomorphic with $SU(2)$. The three parameters can be chosen to be the three Euler angles.

We shall state here, without demonstration, that the property of the finite groups

$$\sum_R D(R) = \sum_R D(SR),$$

which we call the rearrangement theorem, reduces in the case of continuous groups to

$$\int D(R)\ dR = \int D(SR)\ dR$$

where the integrals should be interpreted to mean

$$\int D(R)\ dR = \int D[R(\underline{p})]\ g_L[R(\underline{p})]\ dp_1 dp_2 \cdots dp_n$$

and where $g_L(R)$ is a density function which guarantees that the density of elements R is the same as the density of elements SR. In order to find a suitable $g_L(R)$, we note that:

(a) There is always an arbitrary normalization factor in the integrals; it is then possible to choose arbitrarily the density function in the neighborhood of the identity operation $R = E$

$$g_L(E) = g_0;$$

(b) R takes each volume element V_0 near E to a new volume element V_R near R, and, since the total number of operations is the same in V_R and in V,

$$g_L(R)\ V_R = g_0 V_0$$

that is, $g(R)$ is determined by the contraction of the volume element V_0 as it is moved from E to R;

(c) Since the product

$$R(\underline{p}) \ E(\underline{e}) \ = \ R(\underline{p})$$

implies $\underline{p} = f(\underline{e}, \underline{p})$,

which gives the parameters $\underline{p}$ as functions of the parameters $\underline{e}$, then the change in volume is given by the Jacobian of the transformation. This yields

$$g_L(R) \ = \ g_0 \left\{ \frac{\partial[f(\underline{e},\underline{p})]}{\partial[\underline{e}]} \right\}^{-1}$$

where the derivatives in the Jacobian should be evaluated at the values of the parameter corresponding to the identity.

It should be noted that for non-abelian groups there are <u>two</u> density functions g — one on the left g_L, which is the one defined so far, and one on the right g_R, which guarantees that

$$\int D(R) \ dR = \int D(RS) \ dR$$

$$\int D(R) \ dR = \int D[R(\underline{p})] \ g_R[R(\underline{p})] \ dp_1 \ \cdots \ dp_n$$

$$g_R(R) \ = \ g_0 \left\{ \frac{\partial[f(\underline{p}, \ \underline{e})]}{\partial[\underline{e}]} \right\}^{-1}$$

<u>Examples</u>:

(A) Axial rotation group

$$\varphi_c = \varphi_a + \varphi_b \ , \ f(\varphi_a, \ \varphi_b) = \varphi_a + \varphi_b$$

$$\therefore \ \ \left. \frac{\partial f}{\partial \varphi_a} \right|_{\substack{\varphi_a = 0 \\ \varphi_b = \varphi}} = \left. \frac{\partial f}{\partial \varphi_b} \right|_{\substack{\varphi_b = 0 \\ \varphi_a = \varphi}} = 1$$

$$g_L(\varphi) \cdot = g_R(\varphi) = 1$$

(B) Linear transformations in one dimension.

$$c_1 = b_1 a_1 \qquad\qquad f_1(\underline{a},\underline{b}) = b_1 a_1$$

$$c_2 = b_1 a_2 + b_2 \qquad\qquad f_2(\underline{a},\underline{b}) = b_1 a_2 + b_2$$

$$g_L(a_1, a_2) = \frac{g_0}{a_1^2}$$

$$g_R(a_1, a_2) = \frac{g_0}{a_1}$$

INFINITESIMAL GROUPS AND INFINITESIMAL OPERATORS

The infinitesimal groups are those continuous groups in which all elements are restricted to be infinitesimally close to the identity E, that is, the elements are represented by matrices of the form $R_\epsilon = [E + \epsilon J_R]$, where E is the unit matrix, J_R an arbitrary matrix, and ϵ a small number. In the multiplication rule, all terms in powers of ϵ higher than the first are neglected. Infinitesimal groups <u>are always abelian</u>.

$$[E + \epsilon J_A][E + \epsilon J_B] = E + \epsilon(J_A + J_B) = [E + \epsilon J_B][E + \epsilon J_A]$$

It should be emphasized that $[E + \epsilon J_A]$ and $[E + \epsilon J_B]$ commute, but J_A and J_B <u>do not</u> commute in general. If in an arbitrary continuous group $\underline{e}$ is the set of parameters $\{e_1, e_2 \ldots e_n\}$ corresponding to the identity E, the infinitesimal elements $J_1, J_2 \ldots J_n$ are defined by

$$J_k = \lim_{\alpha \to 0} \frac{R(e_1, e_2 \ldots e_k + \alpha, \ldots, e_n) - E}{\alpha}$$

In the case of transformations in real space, the infinitesimal transformations J_k, or better, the operators

$$I_k = -iP_{Jk}, \text{ [where } P_J f(x) = f(J^{-1}x)]$$

all have a well defined quantum-mechanical physical meaning.

Examples: (A) Group of translations in one dimension.

$$T_a x = x - a$$

$$P_{T_a} f(x) = f(x+a)$$

$$I_T f(x) = -i \lim_{a \to 0} \frac{f(x + a) - f(x)}{a} = -i \frac{df}{dx}$$

In other words, I_T is, in this case, proportional to the momentum operator

$$\hbar I_T = p = -i\hbar \frac{d}{dx}$$

(B) Group of translations in three dimensions.

$$\hbar I_{T_x} = p_x = -i\hbar \frac{\partial}{\partial x}$$

and similarly for I_{T_y} and I_{T_z}.

(C) Group of rotations about the z-axis.

$$I_z f(\varphi) = -i \frac{\partial f}{\partial \varphi}$$

$$\therefore \quad \hbar I_z = -i\hbar \frac{\partial}{\partial \varphi} = L_z$$

The infinitesimal rotation in this case corresponds to the z-component of the angular momentum.

It is important to notice that in order to determine
the representations of a continuous group it is enough to
determine the representations of the infintesimal operators
I_R, since a finite operation $P_R(\alpha)$ can be obtained by a
successive application of I_R a very large number of times:

$$i I_R = \lim_{\epsilon \to 0} \frac{P_R(\epsilon) - P_R(0)}{\epsilon}$$

where $P_R(0)$ corresponds to the identity E.

$$P_R(\epsilon) \approx E + i\epsilon I_R \qquad \epsilon \text{ infinitesimal}$$

$$P_R(\alpha) = \lim_{n \to \infty} (E + i\frac{\alpha}{n} I_R)^n \qquad \alpha \text{ arbitrary}$$

which reduces to

$$P_R(\alpha) = E + i\alpha I_R + \frac{(i\alpha I_R)^2}{2!} + \frac{(i\alpha I_R)^3}{3!} \cdots$$

or, symbolically,

$$P_R(\alpha) = \exp(i\alpha I_R)$$

AXIAL ROTATION GROUP

This is the continuous group of rotations by an angle
φ (clockwise) about the z-axis. It is indicated in the
Schoenflies notation by C_∞ and by ∞ in the international notation. We have seen that it is an abelian group and that
its infinitesimal operator I is

$$I_z = -i\frac{\partial}{\partial \varphi} = iy\frac{\partial}{\partial x} - ix\frac{\partial}{\partial y} = \hbar^{-1} L_z.$$

Since the group is abelian, all irreducible representations
are one-dimensional. The characters (or matrices) are such
that

$$\chi(\varphi_a + \varphi_b) = \chi(\varphi_a) \, \chi(\varphi_b),$$

which requires

$$\chi(\varphi) = e^{c\varphi};$$

but, since χ should be single-valued,

$$\chi(\varphi + 2\pi) = \chi(\varphi);$$

the constant c can thus only take values im where m is an integer:

$$\chi^{(m)}(\varphi) = e^{im\varphi}$$

The orthogonality theorem takes the form

$$\int_0^{2\pi} \chi^{(m)}(\varphi)^* \, \chi^{(m')}(\varphi) d\varphi = 2\pi\delta_{mm'}$$

Since (m) and (-m) are complex-conjugate representations, they are degenerate by time-reversal symmetry and can be considered to form only one representation. The final character table is then

Group $C_\infty(\infty)$

Representation		E	$C(\varphi)$	Functions
A_1	Σ	1	1	z, z^2, x^2+y^2
E_1	Π	2	$2 \cos \varphi$	(x,y), (xz,yz)
E_2	Δ	2	$2 \cos 2\varphi$	(x^2-y^2, xy)

If in addition to being invariant under any rotation about the z-axis the system is also invariant under reflection on any plane passing through the z-axis, the relevant group (a mixed continuous group now) is $C_{\infty V}$ (or ∞m). It can be seen immediately that the group is no longer abelian. The rotations by φ and $(-\varphi)$ form a class while all reflections are in the same class. The character table is now

Group $C_{\infty V}$ (∞m)

Representation		E	$C(\varphi)$	σ_V	Functions
A_1	Σ^+	1	1	1	z, z^2, x^2+y^2
A_2	Σ^-	1	1	-1	$(xy'-x'y)$
E_1	Π	2	$2\cos\varphi$	0	$(x,y);\ (xz,yz)$
E_2	Δ	2	$2\cos 2\varphi$	0	(x^2-y^2, xy)

We can expand the $C_{\infty V}$ group once more by including an inversion center, changing the group now to $D_{\infty h}$ (∞/mm). $D_{\infty h}$ is a direct product group of $C_{\infty V}$ and the inversion group. Its elements are

E	the identity
$C(\varphi)$	rotations by φ about the z-axis
σ_V	reflections on planes passing through the z-axis
i	the inversion
$iC(\varphi)$	inversion-rotations
$i\sigma_V$	rotations by π about axes perpendicular to the z-axis passing through the inversion center.

The character table is given by

Group $D_{\infty h}$ (∞/mm)

Representation	E	$C(\varphi)$	σ_v	i	$iC(\varphi)$	$i\sigma_v$
Σ_g^+	1	1	1	1	1	1
Σ_u^+	1	1	1	-1	-1	-1
Σ_g^-	1	1	-1	1	1	-1
Σ_u^-	1	1	-1	-1	-1	1
Π_g	2	$2\cos\varphi$	0	2	$2\cos\varphi$	0
Π_u	2	$2\cos\varphi$	0	-2	$-2\cos\varphi$	0
Δ_g	2	$2\cos 2\varphi$	0	2	$2\cos 2\varphi$	0
Δ_u	2	$2\cos 2\varphi$	0	-2	$-2\cos 2\varphi$	0

In all these axial groups the notation of the representations is as follows:

(a) Σ, Π, Δ, Φ correspond to $|m| = 0, 1, 2, 3$, respectively, where the capital Greek symbols correspond to the equivalent s, p, d, f sequence of the atomic wave functions.

(b) The superscript $+$ $(-)$ indicates the even (odd) character of the representation with respect to the σ_v reflections.

(c) The g (u) subscript denotes even (odd) character with respect to the inversion.

THE FULL ROTATION GROUP

In what follows we shall indicate a given rotation by $R(\alpha,\underline{n})$, meaning a rotation clockwise by an angle α about the axis $\underline{n}$. The corresponding operator $P_R(\alpha,\underline{n})$ corresponds to a rotation α clockwise of the coordinate axes about $\underline{n}$ or equivalently a rotation α counterclockwise of the point (x,y,z) about $\underline{n}$. As we have seen, the corresponding infinitesimal rotation operator $I_{\underline{n}}$ is given by

$$i I_{\underline{n}} = \lim_{\alpha \to 0} \frac{P_R(\alpha,\underline{n})-E}{\alpha} = \frac{i}{h} L_{\underline{n}};$$

the operator P_R for small angle $\delta\alpha$ is

$$P_R(\delta\alpha,\underline{n}) \cong E + i\,\delta\alpha I_{\underline{n}}$$

and for finite angle α takes the form

$$P_R(\alpha,\underline{n}) = \exp[i\alpha I_{\underline{n}}],$$

where the last expression means symbolically a formal expansion in a series identical to the exponential series.

Consider now the following rotation $P_R(\alpha,\underline{n})$

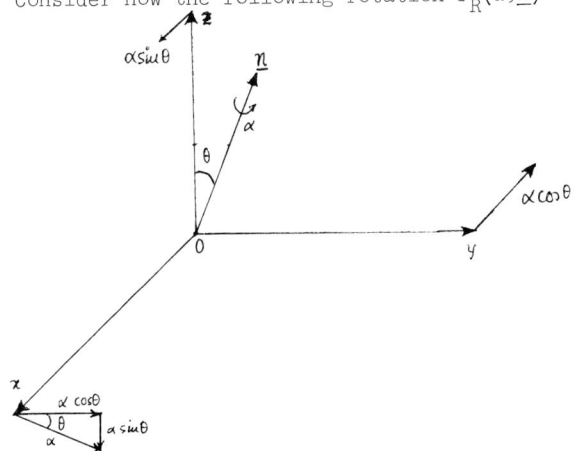

where $\underline{n}$ is in the yz-plane, at an angle θ from z toward y. From the figure it is evident that for small α (neglecting terms in α^2),

$$P_R(\alpha,\underline{n}) = P_R(\alpha \sin \theta,y)\ P_R(\alpha \cos \theta,z)$$

where on the right-hand side y (or z) means $\underline{n}$ parallel to the y (or z) axis. Using the expression for small angles,

$$E + i\alpha I_{\underline{n}} \cong (E + i\alpha \sin \theta I_y)(E + i\alpha \cos \theta I_z),$$

we obtain

$$I_{\underline{n}} = I_y \sin \theta + I_z \cos \theta.$$

This expression can be generalized for a general direction

$$I_{\underline{n}} = kI_x + \ell I_y + nI_z$$

where k, ℓ, m are the three direction cosines of $\underline{n}$.

If we now place a physical body at the origin in the figure and perform on it a rotation $R(\alpha,\underline{n})$, it is evident that this can be accomplished in three steps:

1. Rotate the body by an angle θ counterclockwise about the **x**-axis: $R(-\theta,x)$.

2. Rotate the body by an angle α about the z-axis: $R(\alpha,z)$.

3. Rotate the body again about the x-axis this time by an angle θ clockwise in order to restore $\underline{n}$ to its original position: $R(\theta,x)$.

Consequently,

$$R(\alpha,\underline{n}) = R(\theta,x)\ R(\alpha,z)\ R(-\theta,x)$$

or, since R and P_R are isomorphic,

$$P_R(\alpha,\underline{n}) = P_R(\theta,x)\ P_R(\alpha,z)\ P_R(-\theta,x).$$

This expression is valid for any values of the angles α and θ. It is thus possible to expand it in a power series in α and θ and equate coefficients

$$E + i\alpha[I_y \sin \theta + I_z \cos \theta] + O(\alpha^2)$$

$$= [E + i\theta I_x + O(\theta^2)][E + i\alpha I_z + O(\alpha^2)][E - i\theta I_x + O(\theta^2)]$$

The term proportional to $\alpha\theta$ yields the following equation

$$iI_y = -I_x I_z + I_z I_x$$

In a similar way we can prove that

$$\boxed{I_1 I_2 - I_2 I_1 = iI_3}$$

for $(1, 2, 3)$ equal to (x, y, z) in cyclic order. These are the commutation relations from which all further properties will be derived. It is useful to define I_+ and I_-

$$I_{\pm} = I_x \pm iI_y$$

and the commutator symbol

$$[a,b] \equiv ab - ba.$$

In terms of these, the commutation relations are

$$\boxed{\begin{array}{l} [I_+, I_-] = iI_z \\\\ [I_z, I_+] = I_+ \\\\ [I_z, I_-] = I_- \end{array}}$$

We now look for irreducible representations of the full rotation group such that the basis functions u are also basis functions for the irreducible representations of the axial rotation group about the z axis, that is,

$$P_R(\alpha, z)\, u_m = e^{im\alpha}\, u_m$$

Expansion in powers of α yields

$$\boxed{I_z\, u_m = m\, u_m.}$$

We prove now that if u_m transforms according to the mth representation of the axial rotation group, $I_+ u_m$ transforms according to the $(m + 1)^{\text{th}}$ irreducible representation

$$I_z(I_+ u_m) = [I_+ + I_+ I_z]\, u_m = I_+[E + I_z]\, u_m$$

$$= I_+\, (m + 1)\, u_m = (m + 1)(I_+ u_m). \qquad \text{Q.E.D.}$$

In a similar way $I_- u_m$ transforms according to the $(m - 1)^{\text{th}}$ irreducible representation of the axial rotation group:

$$I_z(I_- u_m) = (m - 1)(I_- u_m).$$

We are interested in the _finite_ representations of the full rotation group. Consequently we choose a number j which is the maximum value that m may take, and impose the condition

$$I_+\, u_j = 0.$$

In order to identify this representation we need now two labels for u: m and j. They are such that

$$I_z\, u_{jm} = m\, u_{jm}$$

$$I_+\, u_{jj} = 0.$$

Since I_- acting on u_{jm} produces a function proportional to $u_{j(m-1)}$, we write

$$u_{j(m-1)} = \alpha_{jm} I_- u_{jm}.$$

Moreover

$$I_+ u_{j(m-1)} = \beta_{jm} u_{jm}.$$

If we write

$$\beta_{jm} = c_{j(m-1)} \alpha_{jm},$$

we can demonstrate that the coefficient c satisfies the difference equation

$$c_{j(m-1)} = c_{jm} + 2m$$

$$c_{jj} = 0 \text{ (boundary condition)}$$

Proof:

$$I_+ u_{j(m-1)} = \underline{c_{j(m-1)} \alpha_{jm} u_{jm}} \quad \text{(by definition of c)}$$

$$I_+ u_{j(m-1)} = \alpha_{jm} I_+ I_- u_{jm} \quad \begin{array}{l}\text{(by replacing } u_{j(m-1)} \\ \text{from formula above)}\end{array}$$

$$= \alpha_{jm}[I_- I_+ + 2I_z] u_{jm} \quad \begin{array}{l}\text{(from commutation re-} \\ \text{lation)}\end{array}$$

$$= \alpha_{jm}[I_- c_{jm} \alpha_{j(m+1)} u_{j(m+1)} + 2m u_{jm}]$$
$$\quad\quad (I_+ \text{ and } I_z \text{ acting on } u_{jm})$$
$$= \alpha_{jm}[c_{jm} u_{jm} + 2m u_{jm}] \quad (I_- \text{ acting on } u_{j(m+1)}$$

$$= \underline{\alpha_{jm}[c_{jm} + 2m] u_{jm}} \quad\quad \text{Q.E.D.}$$

The solution of the difference equation is

$$c_{jm} = A_j - m(m+1)$$

and the application of the boundary condition $c_{jj} = 0$ yields

$$c_{jm} = j(j+1) - m(m+1)$$

To determine α_{jm}, we apply the normalization condition

$$\int u^*_{jm} u_{jm} \, d\tau = 1 \quad \text{for all } j \text{ and } m,$$

and note that

$$\int u^* (I_+ v) \, d\tau = \int (I_- u)^* v \, d\tau$$

-as can be proved from the definitions and the invariance of the integral under rotations.
Thus

$$1 = \int u^*_{jm} u_{jm} \, d\tau = \alpha_{j(m+1)} \int (I_- u_{j(m+1)})^* u_{jm} \, d\tau$$

$$= \alpha_{j(m+1)} \int u^*_{j(m+1)} (I_+ u_{jm}) \, d\tau$$

$$= \alpha_{j(m+1)} c_{jm} \alpha_{j(m+1)} \int u^*_{j(m+1)} u_{j(m+1)} \, d\tau$$

$$= [\alpha_j(m+1)]^2 c_{jm}$$

And finally

$$\boxed{\begin{aligned} I_+ \, u_{jm} &= \sqrt{j(j+1) - m(m+1)} \; u_{j(m+1)} \\ I_- \, u_{jm} &= \sqrt{j(j+1) - m(m-1)} \; u_{j(m-1)} \\ I_z \, u_{jm} &= m \, u_{jm} \end{aligned}}$$

The representations are finite, since

$$-j \leq m \leq j$$

that is, they consist of $(2j + 1)$ functions. Thus far we have assumed m and consequently j to be integers. However if we remove the condition of single-valuedness of the u_{jm} and impose only the condition of restricting ourselves to finite representations, we are left with

$$2j + 1 = \text{integer},$$

or
$$j = 0, \frac{1}{2}, 1, \frac{3}{2}, \ldots, \frac{n}{2}, \ldots$$

For integral j, m is also integral and the representation is single-valued; for half-integral j, m is half-integral and the representations are double-valued, with a change in sign for any rotation by 2π. We shall state without proof that these representations form a complete system; that is, they are all the irreducible representations we can find. They are denoted by the symbol $D^{(j)}(\alpha, \underline{n})$.

Examples:

(A) $j = 0$. Only one function u_{00}.

$$I_+ = I_- = I_z = 0, \quad D^{(0)}(\alpha, \underline{n}) = 1 .$$

(B) $j = \frac{1}{2}$. Two functions $u_{\frac{1}{2}, \frac{1}{2}}$, $u_{\frac{1}{2}, -\frac{1}{2}}$

$$I_z = \frac{1}{2} \begin{pmatrix} 1 & 0 \\ 0 & -1 \end{pmatrix}, \quad I_+ = \begin{pmatrix} 0 & 1 \\ 0 & 0 \end{pmatrix}, \quad I_- = \begin{pmatrix} 0 & 0 \\ 1 & 0 \end{pmatrix} .$$

Define $2I_{x,y,z} \equiv \sigma_{x,y,z}$

$$\sigma_x = \begin{pmatrix} 0 & 1 \\ 1 & 0 \end{pmatrix} , \quad \sigma_y = \begin{pmatrix} 0 & -i \\ i & 0 \end{pmatrix} , \quad \sigma_z = \begin{pmatrix} 1 & 0 \\ 0 & -1 \end{pmatrix}$$

Properties: $\sigma_i^2 = E = \begin{pmatrix} 1 & 0 \\ 0 & 1 \end{pmatrix}$

$$D^{\frac{1}{2}}(\alpha, z) = \exp(i\alpha I_z) = \sum_{n=0}^{\infty} \left(\frac{i}{2}\alpha\right)^n \frac{1}{n!} \sigma_z^n$$

$$= E \cos \frac{\alpha}{2} + i\sigma_z \sin \frac{\alpha}{2} = \begin{pmatrix} e^{i\frac{\alpha}{2}} & 0 \\ 0 & e^{-i\frac{\alpha}{2}} \end{pmatrix}$$

$$D^{\frac{1}{2}}(\beta, y) = \exp(i\beta I_y) = \sum_{n=0}^{\infty} \left(\frac{i}{2}\beta\right)^n \frac{1}{n!} \sigma_y^n$$

$$= E \cos \frac{\beta}{2} + i\sigma_y \sin \frac{\beta}{2} = \begin{pmatrix} \cos \frac{\beta}{2} & \sin \frac{\beta}{2} \\ \sin \frac{\beta}{2} & \cos \frac{\beta}{2} \end{pmatrix}$$

For a general rotation $P_R(\alpha, \underline{n})$ where the value and axis are indicated by the three Euler angles φ, θ, χ, $D^{\frac{1}{2}}(\alpha, \underline{n})$ is given by

$$D^{\frac{1}{2}}(\alpha, \underline{n}) = D^{\frac{1}{2}}(\chi, z) \quad D^{\frac{1}{2}}(\theta, y) \quad D^{\frac{1}{2}}(\varphi, z)$$

$$= \pm \begin{pmatrix} \exp \frac{i}{2}(\chi+\varphi)\cos \frac{\theta}{2} & \exp \frac{i}{2}(\chi-\varphi)\sin \frac{\theta}{2} \\ -\exp \frac{i}{2}(\chi-\varphi)\sin \frac{\theta}{2} & \exp \frac{i}{2}(-\chi-\varphi)\cos \frac{\theta}{2} \end{pmatrix}$$

where the $\pm$ sign is due to the half-integral value of j.

(C) $j = 1$. Three functions u_1, u_0, u_{-1} .

$$I_+ u_1 = 0 \qquad I_+ u_0 = \sqrt{2}\, u_1 \qquad I_+ u_{-1} = \sqrt{2}\, u_0$$

$$I_z u_1 = u_1 \qquad I_z u_0 = 0 \qquad I_z u_{-1} = -u_{-1}$$

$$I_- u_1 = \sqrt{2}\, u_0 \qquad I_- u_0 = \sqrt{2}\, u_{-1} \qquad I_- u_{-1} = 0$$

$$I_+ = \begin{pmatrix} 0 & \sqrt{2} & 0 \\ 0 & 0 & \sqrt{2} \\ 0 & 0 & 0 \end{pmatrix} \qquad I_z = \begin{pmatrix} 1 & 0 & 0 \\ 0 & 0 & 0 \\ 0 & 0 & -1 \end{pmatrix} \qquad I_- = \begin{pmatrix} 0 & 0 & 0 \\ \sqrt{2} & 0 & 0 \\ 0 & \sqrt{2} & 0 \end{pmatrix}$$

Examples of functions transforming like $D^{(1)}$ are the $\ell = 1$ spherical harmonics

$$u_1 = Y_1^{\;1} \propto x + iy$$
$$u_0 = Y_1^{\;0} \propto z$$
$$u_{-1} = Y_1^{\;-1} \propto x - iy$$

DIRECT PRODUCT REPRESENTATIONS $D^{(j)} \times D^{(j')}$

Consider two sets of functions $\{u_{jm}\}$ and $\{v_{j'm'}\}$ which transform according to the $D^{(j)}$ and $D^{(j')}$ representations of the full rotation group, respectively. We are interested in determining how the direct product $\{u_{jm}v_{j'm'}\}$ which transforms according to $D^{(j)} \times D^{(j')}$ can be reduced to its irreducible components:

$$D^{(j)} \times D^{(j')} = \sum_J a_J D^{(J)}$$

We shall prove in a non-rigorous way that

$$a_J = 1, \; |j - j'| \leq J \leq j + j'$$

$$= 0 \text{ , otherwise.}$$

The way of performing the reduction can be summarized as follows:

(1) Form the $(2j + 1)(2j' + 1)$ products $u_{jm}v_{jm'}$ and classify them according to the irreducible representa-

tions M of the axial rotation group about the z-axis. Remember that $u_{jm}v_{j'm'}$ transforms according to $M = m + m'$.

(2) Select the vector

$$U_{(j+j')(j+j')} = u_{jj}v_{j'j'}$$

(3) Generate by means of the operator I_- and the proper coefficients, the $(2j + 2j' + 1)$ vectors $U_{(j+j')M}$ which transform according to $D^{(j+j')}$.

(4) Orthogonalize all other vectors to the already determined U_{JM}. This can be accomplished by taking linear combinations of $u_{jm}v_{jm'}$ in which vectors with different $M = m + m'$ do not mix.

(5) Take from the remaining vectors the one with highest M, call that M the new J, and call that vector U_{JJ}.

(6) Generate by means of I_- the remaining 2J vectors U_{JM} of the D^J representation.

(7) Repeat steps (4) to (7) until the whole space has been exhausted.

The vectors of the new irreducible representations U_{JM} can be expressed as

$$U_{JM} = \sum_{jj'mm'} (jj'mm' | JM) \, u_{jm}v_{j'm'}$$

The coefficients $(jj'mm' | JM)$ are called the Wigner or Clebsch-Gordan coefficients; they are zero unless $m+m'=M$. From the method proposed above we see immediately that in the set $\{u_{jm}v_{j'm'}\}$

(i) there is only one function with

$$M \equiv m + m' = (j + j')$$

and similarly for $M = -(j + j')$;

(ii) there are two functions with

$$M = (j + j' -1) \quad \text{or} \quad M = -(j + j' -1)$$

unless j and/or j' are zero;

(iii) there are three functions with

$$M = (j + j' - 2) \text{ or } M = -(j + j' - 1)$$

unless j and/or j' are zero or one;

(iv) values of M such that

$$- | j - j' | \leq M \leq | j - j' |$$

are contained $(2j + 1)$ times if $j \leq j'$ or $(2j' + 1)$ times if $j' < j$.

Consequently, straightforward counting yields

$$D^{(j)} \times D^{(j')} = D^{(j+j')} + D^{(j+j'-1)} + \ldots + D^{(| j-j' |)}$$

Example: Let us consider $(u_1 \; u_0 \; u_{-1})$ and $(v_1 \; v_0 \; v_{-1})$ which both transform according to $D^{(1)}$. Our formula gives for the direct product

$$D^{(1)} \times D^{(1)} = D^{(2)} + D^{(1)} + D^{(0)}$$

We shall determine the 9 functions U_{JM} by following the procedure outlined before:

(a) Classification of the direct product functions.

M = 2	$u_1 v_1$	
M = 1	$u_1 v_0$	$u_0 v_1$
M = 0	$u_1 v_{-1}$	$u_0 v_0$	$u_{-1} v_1$
M = -1		$u_0 v_{-1}$	$u_{-1} v_0$
M = -2			$u_{-1} v_{-1}$

(b) Determination of $D^{(2)}$

$$\boxed{U_{22} = u_1 v_1}$$

$$I_- U_{22} = 2U_{21} = I_-(u_1v_1) = (I_-u_1)v_1 + u_1(I_-v_1) =$$

$$= \sqrt{2} \, u_0v_1 + \sqrt{2} \, u_1v_0$$

$$\therefore \quad \boxed{U_{21} = \frac{1}{\sqrt{2}} \, (\, u_0v_1 + u_1v_0)}$$

$$I_- U_{21} = \sqrt{6} \, U_{20} = \frac{1}{\sqrt{2}} \, [(I_-u_0)v_1 + u_0(I_-v_1) + (I_-u_1)v_0 + u(I_-v_0)]$$

$$= \frac{1}{\sqrt{2}} \, [\sqrt{2} \, u_{-1}v_1 + \sqrt{2} \, u_0v_0 + \sqrt{2} \, u_0v_0 + \sqrt{2} \, u_1v_{-1}$$

$$\therefore \quad \boxed{U_{20} = \frac{1}{\sqrt{6}} \, [u_{-1}v_1 + 2u_0v_0 + u_1v_{-1}]}$$

Similarly
$$\boxed{U_{2,-1} = \frac{1}{\sqrt{2}} \, (u_0v_{-1} + u_{-1}v_0)}$$

$$\boxed{U_{2,-2} = u_{-1}v_{-1}}$$

(c) Orthogonalization of the M = 1 functions to the U_{21} function

$$\boxed{U_{11} = \frac{1}{\sqrt{2}} \, (u_0v_1 - u_1v_0)}$$

(d) Determination of $D^{(1)}$

$$I_- U_{11} = \sqrt{2}U_{10} = \frac{1}{\sqrt{2}} \, [\sqrt{2} \, u_1v_1 + \sqrt{2} \, u_0v_0 - \sqrt{2} \, u_0v_0 - \sqrt{2} \, u_1v_{-1}]$$

$$\boxed{U_{10} = \frac{1}{\sqrt{2}} \, [u_{-1}v_1 - u_1v_{-1}]}$$

and

$$U_{1,-1} = \frac{1}{\sqrt{2}} (u_{-1}v_0 - u_0v_{-1})$$

(e) Orthogonalization of the M = 0 function to the U_{20} and U_{10} functions, which determines U_{00}

$$U_{00} = \frac{1}{\sqrt{3}} (u_1v_{-1} - u_0v_0 + u_{-1}v_1)$$

(f) The Clebsch-Gordan coefficients. By writing

$$U_{JM} = \sum_{jj'mm'} (jj'mm' \mid JM) u_{jm}v_{j'm'}$$

we see that in this example we have determined:

$(1111 \mid 22) = 1$

$(1101 \mid 21) = \frac{1}{\sqrt{2}}$

$(11\text{-}11 \mid 20) = \frac{1}{\sqrt{6}}$

$(1100 \mid 20) = \frac{2}{\sqrt{6}}$

$(1101 \mid 11) = \frac{1}{\sqrt{2}}$

$(1100 \mid 00) = \frac{1}{\sqrt{3}}$

etc.

THE ROTATION-INVERSION GROUP

The full rotation group, which contains all proper rotations, can be expanded to include the inversion and improper rotations. We can easily do this by taking the direct product between the full rotation group and the group of the inversion. The resulting group is known as the rotation-inversion group or the full orthogonal group O(3).

The representations of O(3) are in direct correspondence with the representations of the proper-rotation group, there being two irreducible representations of O(3) for each one of $O^+(3)$. The even representation is such that

$$D_+^{(\ell)}(iR) = D_+^{(\ell)}(R)$$

while the odd representation satisfies

$$D_-^{(\ell)}(iR) = -D_-^{(\ell)}(R)$$

If we consider a single spinless particle, the parity of the functions transforming according to $D^{(\ell)}$ is given by $(-1)^\ell$. If the system consists of n spinless particles, the parity is given by

$$\prod_{k=1}^{n} (-1)^{\ell_k}$$

where ℓ_k is the orbital angular momentum of the kth particle.

When spin is introduced and j takes half-integer values, the ambiguity of sign in the representation matrices obscures the simple distinction between odd and even functions, and, although parity is still a significant quantum number, its usefulness as a tool in determining selection rules is substantially decreased.

INDEX